Legal

ISBN: 978-1-7321877-2-6

This publication is designed to provide accurate and authoritative information in regard to the subject matter covered. It is sold with the understanding that neither the author nor the publisher is engaged in rendering legal, accounting, or other professional service. If legal advice or other expert assistance is required, the services of a competent professional person should be sought.

--From a Declaration of Principles Jointly Adopted by a Committee of the American Bar Association and a Committee of Publishers and Associations.

Thank You

This book is dedicated to my family, the original and strongest supporters of my technology.

Introduction

The 3nd volume combines micro and macroeconomic analysis, deal flow, banking, organization behavior and venture capital.

Matthew L. Myers presenting Upgrade It!® at SF New Tech®.

Table of Contents

Precursor

My process of examining deal flow and M&A activity started while riding the bus to and from work in Honolulu, Hawai'i. Many of the themes have remained the same: SAAS and Cleantech. However, new segments have emerged as well: Equipment as a Service (EAAS), Fintech and Insurtech. The vision of a solar and renewable future provided a strong market and value proposition at the time. However, the economies of scale and the limits of power purchase agreements with utilities restricted the technologies larger adoption in the late 2000s.

Several years later while riding the BART in San Francisco it was clear that technology was reaching an inflection point – ridership and traffic were exploding in the Bay Area; and Los Angeles and San Diego were beginning to emerge as hot spots of innovation. Investments that initially shifted away from Cleantech and other renewable innovations now are beginning to circle-back and re-emerge as sectors of capital fundraising. The infrastructure and housing did not keep pace with the influx of new software firms. In many instances the condition of the roads, schools and bridges actually became worse.

What is clear from examining technology and banking over a ten (10) year period is that consumer and corporate behavior has changed very little. Traffic and energy use is higher than ever. The infrastructure in the United States and the world has not kept up with the influx of new start-ups and incumbents into California, New York, Texas and elsewhere.

The type of visionary technology that emerged during Bell Labs® heyday has *not* materialized. Most software technology is similar to its precursors focusing on process improvements, data organization, algorithms or AI.

A new energy source and means of transporting and housing large segments of the population more cost effectively or efficiently has not arisen. In essence, new technology innovation has largely consisted of refinements to existing technologies (i.e. project management, word processing, data organization and otherwise). It has produced an electric car, LED light bulb and LEED housing. An alternative to the car, light bulb or housing has not surfaced.

Energy, transportation and housing efficiency and innovation has become inverted and is actually declining. Cities like San Francisco, New York, Los Angeles have been unable to alleviate traffic, congestion and provide adequate housing for their workforce and residents.

From a traffic management perspective adding more lanes is not the solution (even tunnels underground is not the answer). The cars still have to merge with existing traffic even if they are entering a lane from a sub-surface elevator. The net effect is still the same. There is not physically enough room for merging and right-of-ways.

Why is this the case?

The exit lanes off a freeway into a city like San Francisco neck down from seven lanes to two and then even one. In essence, the streets in San Francisco would all have to be widened to fourteen lanes (seven on each side) in the urban center to avoid car traffic back-up that is taking the down ramp into the city.

As it is physically impossible to widen city streets to accommodate this change in most metropolitan areas, the more practical alternative is to add more high-speed mass transit capacity (bus and rail) in combination with motor vehicle traffic.

The larger issue is that the United States is still one of the only countries without high-speed rail (bullet trains). The Hyperloop One® provides the potential for groundbreaking change in the area of multiple occupant mass-transit, *not* underground car tunnels.

Housing innovations are desperately needed. Concrete, glass, wood and metal have not been supplanted by new material advances.

Scientists and engineers have *not* produced a *new* groundbreaking technology capable of providing energy in cost effective and efficient manner.

However, wave energy has the potential to deliver exceptional volume, especially in coastal urban areas, where many large population centers are. Geothermal has the capability of addressing a portion of the needs of inland urban areas.

Innovation is not just software and hardware. To truly provide a value proposition that is exceptional it has to rise to the level of originality not seen before: a new light-based data storage, or in the physical arena an invention like Velcro or the bicycle chain. Many of these ideas appear in good science fiction writing; i.e. *Dick Tracy's* watch manifests itself years later; or the stories found in *A Man Called X*.

The start-ups and technology that I met with, interviewed and conducted due diligence on are all trying to build new brands and start-ups. They are to be strongly commended for this effort and the accomplishments that they have made. However, they have not really provided solutions to the truly large-scale problems that society faces and they can.

U.S. Venture Capital Ten-Year Comparison Report

The focus of investing has changed over a twelve-year (12) period from approximately 2007 to 2019 with fewer larger-sized dollar deals at the high end of the spectrum and a shift in the overall number of transactions (volume) of the marketplace. Deal flow, which is always a changing target, has shifted for investors and M&A (in many cases) away from smaller seed and angel-backed companies into single heavily branded companies with multiple prior rounds, including Uber® and Airbnb®.

Venture Capital has moved from Cleantech through Adtech, Artificial Intelligence (AI), Virtual Realty (VR), Cybersecurity, Equipment as a Service (EAAS), Gaming, Smart Logistics, Supply Chain Management, Agtech and now Insurtech and Fintech; including Blockchain (digital ledgers), Bitcoin and now Initial Coin Offerings (ICO); or offerings for crypto currencies and blockchain. Many of these securities' offerings are not legitimate, relying on a .pdf document or email and should be avoided at all costs. The U.S. Securities and Exchange Commission (SEC) are not providing approval to a large majority of the ICO offerings issued as asset offerings. Investors should use common sense. It takes a lot of preparation for an Initial Public Offering (IPO), and due diligence and regulatory filings provide the underpinning for a sound investment.

Smart Logistics and Supply Chain Management (SCM)

Smart Logistics and Supply Chain Management (SCM) have emerged as stand-alone categories of venture capital investment. Fleet management and warehousing are areas where sensors, lidar and smart containers are changing the way companies do business with industry. There were approximately 57 deals in 2017 in the SCM and Smart Logistics groups.[1] These products include software as a service (SAAS), platform as a service (PAAS) combined with big data analytics in the cloud and sensor based tracking. These issues are revolutionizing the way companies' track and ship inventory and merchandise. Tracking mechanisms are through Internet of Things (IOT), which combine cell and/or fiber-based tracking with satellite-based pinging to monitor car, truck, ships, containers and entire fleets; and even field service workers.

---

[1]Confidential M&A deal flow meeting. 2018.

Top buyers in the sector include Trimble®, WiseTech® and Descartes®. WiseTech Logistics® is a leader in smart logistics. Some of the M&A deals in 2017 included Verizon® acquiring Skyward®; Walmart's® acquisition of Parcel®, Target® acquired Shipt®, MacroPoint® was acquired by Descartes®, Pitney Bowes® picked up Logistics® and in a massive deal SoftBank® and Alphabet® acquired Manbang Group® (a truck sharing and calling app) for $2 billion.

An interesting trend according to the team at Connected Holdings® is the emergence of the use of blockchain ledgers and infrastructure to track shipments. The integration of blockchain in transportation will continue to be a category of M&A acquisition, combined with ultrasonic cargo sensors, IOT and the usage of satellite tracking. In 2019 and beyond there will continue to be unusual technologies that emerge that are targets of corporate acquisition.

Blockchain, Bitcoin and Initial Coin Offerings

Bitcoin, in many cases, requires exceptional amounts of electricity and specialized video cards and processors to mine (solving riddles and puzzles) and unlock the coins. These virtual currencies are highly volatile and experience heavy fluctuations in valuation. Much of the value of the coins is assigned by speculation and the electrical cost of generating the coins. Blockchain ledgers and newer exchanges including Coinbase® can be used to trade the currencies once they have been mined. One of the original blockchain ledgers was used for Bitcoin transactions.

Any investment offerings in virtual currencies should include the following: prior written approval of the U.S. Securities and Exchange Commission (SEC); the consent of the U.S. state they are operating in; and the sign-off of other branches of the federal government. Investors should also understand any international regulations, including those surrounding wire transfers from countries flagged by the United States as prohibited persons, entities, agents, companies and/or countries.

One of the earliest Blockchain and Bitcoin investment briefings that I attended was held at the Federal Reserve Bank of San Francisco with presentations by Ripple®, Bank of the West® and Visa®. It was clear during this presentation that the technology of Blockchain in particular has the potential to be used not only in banking and venture capital, but also for other non-banking applications.

Criticisms of blockchain ledgers have included the following: it is a thirty (30) year old technology, with the code essentially being a cryptographically signed #hashtag that links a series of documents.

Hedge Funds and Blockchain

One of the most important considerations of blockchain and bitcoin analysis is whether or not it is equity versus providing utility (i.e. Ripple®). Most established firms and investors are concerned with counterpart risk on centralized exchanges; and handling currencies not audited and those with no controls (SOC1 and/or SOC2).

In the bitcoin exchange markets, ledger custodians have emerged as one of the new management firms. These firms advertise the ability to securely hold blockchain ledgers. These firms take snapshots of digital ledgers as custodians of clients' bitcoin holdings. Bitcoin holders have received some regulatory SEC relief in the following ruling: that bitcoin and etherium are not securities. This ruling has increased the buying and selling of bitcoin, which has necessitated an increased need for cold storage, which utilizes broken shards and multiple keys to access and distribute bitcoin and other cryptocurrencies. The record keeping and encryption includes private keys that are broken into digital shards (portions of a record). The shards are distributed to up to 5 different employees, whose identities must be verified through fingerprint and/or retinal verification. It is very difficult to compromise this type of system. The hacker would have to compromise the actual wallet provider, which is walled off through this fragmented system.

However, if a hacker were to compromise this system, once the bitcoin or etherium are gone (as examples), they are very difficult to track. Regulators are still attempting to track down the Mt. Gox exchange loses through a chain analysis-tracing firm. For this reason cyber security insurance is available provides the following: 1st-party loss to 3rd-party vendor breach cyber policies ranging from \$2,500 to \$15,000 for \$1million for one year of coverage. For this asset class (i.e. bitcoin) \$3,500 is the norm and average for the cost of insurance. There is also available separate client insurance for malware protection and denial of service. From an international operational perspective never use anything other than an authorized hotspot to ensure that spear fishing and/or phishing doesn't occur, which allows unauthorized users to hack into phones of CFO and/or CEO.

Due diligence is crucial for service providers from this asset class. For investment firms investing in this asset class there are two types of providers:

    a. Fund SEC registered advisor; and

b. Exempt reporting advisor clients - not registered with you.

As with any 3rd party providers exposing an investment firm to risk, due diligence must include undergoing an audit. Bitcoin was never envisioned originally as a digital asset and pricing for these types of assets are now determined through a counterparty. Dropbox or Google should not be used as a storage facility/solution due to compliance and security issues.

From a regulatory perspective Rule 408 - IRA Sub shares in S Corp, collectibles and one other section provide guidance on where investment should and/or may not be engaged in with respect to Individual Retirement Accounts (IRA).

In general, if you are raising a fund and shift the strategy, this is viewed by institutional investors as a weakness or a sign of a shift in strategy. This is particularly critical for family offices where venture holdings in Fintech and digital payments are involved, and where an independent sponsor backing the fund is necessary.

Family office investments will expect to see less liquidity to minimize tax impacts as a part of 10-year windows of investment, regardless of any ability to show good returns.

Conclusions

Bitcoin is more volatile and subject to extreme price fluctuations. However, blockchain ledgers have the potential to provide the underlying infrastructure for hedge funds and other aggregated venture holdings. The technology has the ability to provide greater structure to asset classes that typically are highly illiquid, particularly hedge funds. Blockchain can provide benchmarking measures that enable funds to provide an alternative means of comparing fund to fund (i.e. Funds I–III) and across different positions based on the amount of assets under management. This has the potential to reduce risk across managers and holdings.

Hedge Fund Platform

One entrepreneur that I interviewed has developed a private equity and hedge fund listing platform.[2] Capital providers and/or other users setup an account. This includes private clients and other independent sponsors. The fee for posting a deal is $750 dollars. Clients may also purchase a package of deal-posting opportunities for $3,000, which includes membership to the firm's proprietary platform (a one year subscription), which is may be under development. The corporation uses social media, sponsored conferences and other traditional advertising to promote these deals. Individual memberships for non-hedge fund members are $500 and $450 to connect to other users and funds.

---

[2]Confidential discussion 6/22/18.

For hedge funds, membership is $10,000 per year, which includes 10 meetings, along with 3 events included as a part of the package for free. Additional fees include $5,000 to host a table and present views on a particular asset class and sector. The site includes a standard non-disclosure agreement to ensure confidentiality. This site and platform is still growing and is currently at a very early stage. Membership has included a family office in India that has posted a deal on this site.

In my visit with the principles in New York it was clear that they had not accurately represented the corporation on two previous occasions, during one telephone communication and one investment presentation. The problem in this case, as is with many startups and family-office groups is that you must deal with the determinant agents; or the entities that actually make the decisions, i.e. the parents, the founders and/or the board. In this case when I finally met the founder, she informed me that their platform was on hold, they were not raising funds and were focusing primarily on their hedge fund events. Communicating this information to me directly initially could have saved a lot of time.

Conclusions

It is imperative that a manager always meets with the funding principles of any hedge, venture or alternative asset fund. This is the guidance from attorneys and lawyers in the family office and hedge fund arena. It is worth spending extra time ensuring that you are going to meet with the funds CEO or founder. Otherwise your plane, rail or car trip will be a wasted effort.

Why spend the time and resources to take the rail, subway or drive to a location if the determinant agents aren't meeting you half way (i.e. taking the time to introduce themselves in person)?

Your visit will become a site-visit then, requiring you to schedule other business meetings and/or capitalize on the stay as a learning experience. In my case I documented the condition of the subway with a series of photographs, set up an alternative investor meeting and went to a museum to look at the samurai swords.

Fintech and Insurtech

On June 6, 2018 I attended a Fintech event held at NASDAQ® headquarters in Times Square, New York, NY. The panel included three Fintech companies and one former Securities and Exchange Commission (SEC) advisor and attorney. At the end of the discussion I asked the panelists if they were aware of the *Fiduciary Rule*?[3]

This rule was commonplace knowledge in banks and investment services circles and was enacted to ensure that investors are protected from financial advisors' investment advice by safeguarding that those advisors act in the best interests of their clients not themselves.

---

[3]The *Fiduciary Rule* has subsequently been formally repealed.

Matthew L. Myers on-route to NASDAQ® Fintech event.

Only the former regulator from the SEC could answer the question regarding the rule. None of the other panelists were able to describe it in any way. For companies developing investment SAAS and mobile apps it is very troubling that they were not knowledgeable regarding regulatory requirements.

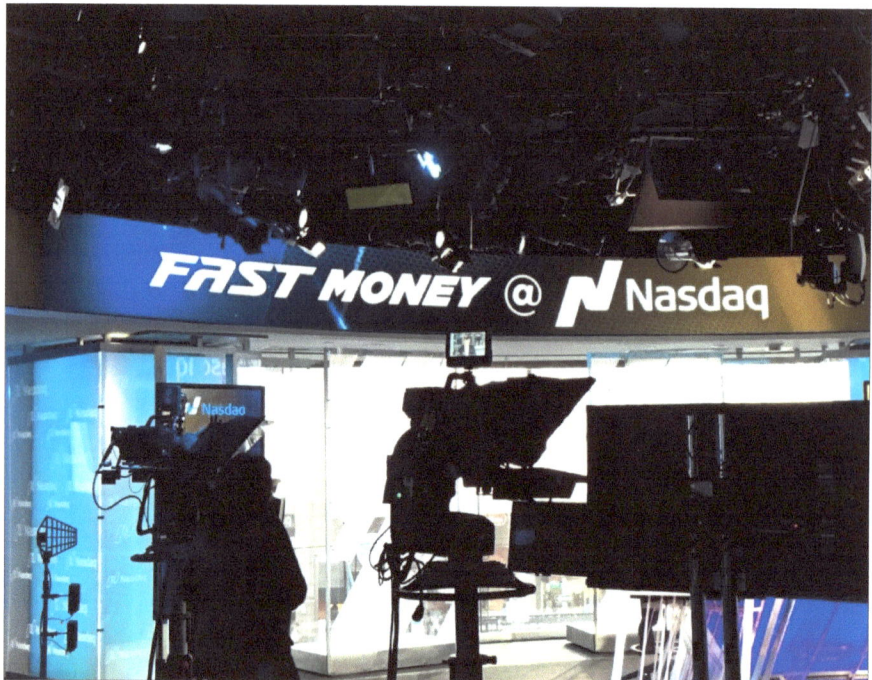

Matthew L. Myers invited onto the set of Fast Money® at NASDAQ®.

Most banks after the financial crisis have been required as a part of the post crash regulatory environment to implement rules regarding federal regulations, including but not limited to, Sarbanes-Oxley, Dodd Frank, Reg W (capital asset allocation thresholds), and Office of Foreign Assets Control (OFAC) constraints. I have provided oversight on developing the code and software integration to ensure that wire transfers and other corporate venture capital (CVC) transactions are in compliance with OFAC and Reg W; (and other internal rules).

The completion required a yearlong audit with a review of data center operations throughout the United States and Europe. An app developer engaging in the development of a finance platform, like a Fintech or Insurtech startup with international transactions must comply with suspicious logins, including multiple sign-ons from difference internet protocols (IP) addresses and with prohibited foreign actors that are flagged in the system.

If a new startup failed to implement these controls they could be liable for U.S. sanctions for conducting business with a prohibited individual, agent, corporation, or nation state (i.e. person from North Korea attempts to open an account within in their platform).

The same is true for other domestic regulations (proposed and implemented). Their platforms should ensure that clients are advised of each of the individual company's procedures for safeguarding investors' funds within their respective platforms. The fact that these companies did not have prior knowledge about these regulations raises a red flag.

Conclusions

Fintech and insurtech are heavily regulated industries. It is unrealistic to expect a decades long securities and regulatory framework to disappear and/or to drastically change. Building new technologies is relatively easy in relationship to ensuring that regulatory requirements are properly met. Investment due diligence should include a thorough examination of front and back-end technologies to ensure that no securities laws are being broken.

Reflections

When I attended hedge fund, bitcoin and fintech presentations in New York City it was clear from walking around the city in advance that the purported benefits (multiplier effect – new jobs and businesses) that these startups were promising had not translated into substantial economic benefits for the city: there were more boarded up businesses and empty store fronts in NYC than on my previous visits. Whole city blocks were empty and vacant. In essence, the startups and funds disruption directly and indirectly had caused large parts of the city to become vacant without providing revenue and benefits for essential repairs to NYC's streets, bridges and subways. The multiplier effect of these companies was inverted: new restaurants, businesses and services were not replacing the empty storefronts.

During the summer months the subway, the cars and transit system were in such bad shape that the air conditioning was broken on many of the cars, delays were frequent and residents were so frustrated that they were banging on the side of the rail cars on a packed platform that was completely full. Once you boarded the train it was standing room only. Arguments and shouting were commonplace. I was thankful that a fight didn't break out.

The streets above (after exiting the subway) were so gridlocked with ride-sharing cars and delivery trucks delivering Amazon® and other on-line packages (the packaging was visible on the street while walking past them being unloaded) that traffic was at a complete standstill. It was more efficient and faster to walk across town in the sweltering 100-degree NYC heat than sit in an on-demand vehicle or taxi.

Why waste your time and money?

The promise of a better future from technology was definitively *not there*. It was the most wasteful, inefficient, least cost-effective and selfish conduct that I have ever observed: Millions of hours of lost productivity, gallons of squandered gasoline and watts of electricity; and broken tracks and AC systems on the rail.

When I met with some of the technology companies after struggling to navigate the streets and subway they were some of the most boring, unoriginal and uninteresting people I have ever encountered. When I ask them: "Have you ever read *x*?"; or "Have you been to the following museum, factory, design center?" It is always "No".

It was clear that to solve these transportation and business inefficiencies it would be necessary to look outside of the *myopically* focused technology community that seemed incapable of original thought and basic problem solving.

This is a group of people that in the classic childhood test of adaptability: if you placed a log in a path in the woods they would be unable to figure out how to get around it.

Technologist are not alone: This is type of excess, inefficiency and lack of inventive thinking is everywhere, at the local landfill reclamation center I have collected enough free masonry and exterior paint (partially used) to cover an entire basement and enormous barn three times with brand new coatings.

On the road that I run on I have collected enough trash from bottles, cans and trash to fill an entire extended cab truck twice. All of this metal and glass needs a marketplace and corresponding new product lines to utilize these materials.

Surplus, chaos and disorganization are everywhere. It is just surprising that new entrepreneurial endeavors don't seek out to tackle these larger problems more systematically. They are missing out on huge market opportunities.

Startups should be commended for their effort to build new products and services. However, they need to begin to figure out how to solve larger problems with technology, not just another consumer product service or ride-hailing app.

New Investment Frontiers

Market segments grow and contract over time. One aspect of investing remains the same. Investors seek to deploy capital efficiently. They will seek out as many new sectors to disrupt as they possibly can. Startups meeting basic investment criteria can excel with the following metrics: monthly recurring revenue (MRR) of $25K at the seed stage, $100K in MRR at the pre-Series-A or Series A stage, providing an employee stock option plan (ESOP) of 10% of operating costs or higher, keeping their cap tables clean and removing inefficiencies in their deployment of capital and bias from their platforms.[4]

---

[4]These metrics are not hard and fast rules. Investors with an appetite for risk may invest in companies with lower MRR, based on higher user growth or other metrics.

Smart Plumbing

A start-up that I examined falls under the research & development (R&D) arm of Samsung® Korea. The firm is a part of Samsung Research America® (SRA) and specifically Samsung® NEXT. The enterprise has made an agreement to transfer 16% of the business for $1,000,000 for the hardware and other IP that the company has developed into a C-Corp. One of the provisions of the successful transfer is that their start-up must raise $5 million in order for SRA to spin the corporation off the start-up and release 100% ownership from SRA as a subsidiary and indirectly from Samsung® Korea.

The two founders come from a start-up background, one having developed a water and wastewater along with energy monitoring SAAS for building owners; and the other a food waste reprocessing program serving Stop&Shop® and Kroger®.

The brand has developed a water flow with meme sensor, having originally attempted Bluetooth and other connectivity hardware with less success. They hired a firmware person to resolve this process and included a firmware machine learning procedure as a part of this process.

The start-up uses LORA Wan – to send small amounts of data long distances. In essence, it utilizes a gateway as an interface to control and regulate apartments. For example, if an apartment complex has 500 units, it requires 50 gateways. The gateway utilizes a packet forward feature to function. The first installations using Bluetooth didn't work.

The firm sends out hardware exterior design instructions to Stratus to design the sensor's casing. The start-up has sent out small iterations of the design, which are 3D printed. The device has had 3 design iterations. The first one was ANT enabled – with a zip tie. As a part of the corporation's Series A, are pre-orders (with contracts) for 18,000 sensors. The endeavor intends to spin out from SRA and start making sensors from 3rd party providers. The casing isn't brittle and doesn't break. Condensation is not an issue. The application is intended for maintenance managers, not plumbers. It can be installed under kitchen sinks in bathrooms and with outdoor irrigation systems.

After it has been installed it is a tool that can diagnose exactly which unit is leaking. The firm's IRR analysis is exceptional and the tool typically indicates that approximately 20% of units are leaking. It has the potential to significantly reduce water usage where there is a flat water rate of $1,100 (as an example) per year. Fifth Wall and Rudin Management are an example of a typical apartment complex and client.

The start-up is an extensive user of 3rd party suppliers, including CircuitHub and Techtronic in Arkansas. Both charge a fee for their services requiring users to pay for each board and sensor, with a sliding scale for high orders. Techtronic (in particular) has the ability to easily move its on-demand manufacturing to Mexico and Asia. The *Made in the USA* label is nice to have and at this juncture the economics work and are viable. Over the long-term, price may dictate suppliers and the corporation may elect to shift manufacturing to Mexico or Asia.

The organization's intent is to never have inventory and drop-ship to customers. They also want customers to purchase packages of 25 or 50 units from the suppliers. This empowers the building owner to do the installations on their own time.

The startup's data and firmware are their long-term focus. Distribution will be completed by HD Supply, a spin out from Home Depot®, that is the largest distributor of maintenance supplies to contractors and is a direct supplier to commercial buildings. It has the capability of handling the sale and distribution of 50,000 sensors.

The firm seeks to train building managers and facility staff. They are not going after the consumer market. The direct-to-consumer market would only be viable if a 3rd party business partner wanted to address this segment. Instead the enterprise is seeking large commercial properties with 5,000 to 10,000 units as a part of their housing portfolio (i.e. Greystar, as an example).

They are also seeking to tackle the affordable housing market and companies that have up to 55,000 affordable-housing apartments in their roster. In this sector they are seeking to be integrated into the LEED certification for water management and conservation in the housing sector. This resolution doesn't require cutting into the pipes in every apartment and offers a solution to where in-line water monitoring wasn't feasible.

They have partnered with Enterprise Green Communities, a 501(c)3 operating out of NYC. This not-for-profit along with the Housing Preservation & Development agency has made water monitoring a required component of their funding and participation. As a consequence, the affordable housing market represents a huge opportunity for the start-up.

The start-up can provide 15% to 20% in water savings (on average) and up to 30% to 40% savings in problematic areas with its technology. For many building owners monitoring leaks is an initial focus area and the emphasis is writing them into the lease renewals.

ARGO Insurance® is capitalizing $1 million as a part of the companies Series A investment. Samsung NEXT® will also own 16% equity via warrants in the post Series A. In order to receive authorization for the technology transfer to occur from Samsung® Korea the company needs to raise at least $3 million as a part of this round. This is a major deal breaker in terms of a fiscal outlay.

In examining the probability involved with not achieving this revenue threshold the asset ceiling limit is simply too great to commit to a formal investment. The deal threshold in this case is acting like a lien on a car or house and carries too much risk. There are some investors that might be willing to take on this speculation, but from a personal perspective I am not. However, the start-up still has a strong value proposition and the technology that it has developed is very viable and solid. It is worth revisiting this business over the long-term.

A $1 million deal in the entrepreneurial endeavor will convert to 6% to 7% ownership (i.e. if an individual invests $5 million that person will own 40%). The business is seeking funds in the $500K to $1 million range. The remaining shares are included as a part of employee equity.

This business uses machine learning and is able to map every faucet and outlet because they have a unique signature. A demonstration of the algorithm was provided at the demo.

The start-up has good customer IRR. Typical cost breakdowns are as follows: $100 per senior and $10,000 per average building, with a required gateway that costs $3,000. Installation of the units (en masse) costs $1,400.

Most of the time installing the units does not require a plumber (just equipment monitoring). The hardware by identifying leaks and having maintenance teams actively addressing has resulted in an average 15% water cost reduction. If a building is experiencing $10,000 in water losses this is offset by water costs going up 6% percent per year. The typical payback period is under a year. Eventually the start-up wants to be a full service provider for buildings and their managers, charging people for water through their platform.

The platform as a service will be particularly valuable in assisted-living facilities. This is due to the algorithms' ability to showcase energy efficiency data and patterns of a typical unit and the overall building. In some buildings the tenants wake up earlier, creating an energy efficiency play.

Conclusions

The startup's endeavor has tremendous value for water conservation initiatives. What is *not* addressed (and what was and is *not* the represented focus of the entrepreneurial endeavor) as a part of the firm's value proposition is the need for more housing units that incorporate new groundbreaking technologies that are strong, light, cost effective and require less energy to build than traditional concrete, glass, lumber and metal (i.e. a new honeycomb substrate like carbon fiber that has not been invented yet). New innovative materials combined with water and energy conservation measures are what are truly needed for cities.

Innovative technology is needed to more efficiently remove lead paint and other hazardous materials in older affordable and market priced units. Many of these units lack adequate heat and running hot water to begin with. The corporation is not addressing some of the more systemic problems that urban and rural communities face in terms of housing. A site visit to older affordable and high-end residential units would be instructive.

From an investment perspective: several corporate umbrellas cloud the conglomerate's IP ownership. An investment in this company is only warranted above the $3 million dollar threshold where the equity is fully transferred to the new ownership. Below $3 million dollars investment is a non-starter.

Smart Car Logistics

This startup provides energy efficiency statistics on car battery usage and other internal diagnostics, in addition to aggregated driver patterns. This is done through a strong visualization of the numbers, creating an eagle-eye view of driver usage.

The corporation wants to shift its revenue focus from large research projects and reports for Johnson Controls® and Ford® ($4 million in revenue, which didn't scale due to its custom nature) to visualizations, licensing and specific reporting produced on demand.

The startup's contract with Johnson Controls® utilizes an algorithm to map the life of batteries across multiple car manufacturers and fleets; essentially buying what they have done. The agreement with Johnson Controls® allows the firm to outsource their technology as long as they don't sell to a direct competitor. The licensing of the large-scale battery project from Johnson Controls® (i.e. battery algorithm – failure and life prediction) and Ford® projects and enterprise licensing have been the start-up's primary focus to date.

The firm uses Amazon® web services and wants sliced facts and enterprise license and reporting to be its major focus. The revenue stream is as follows: flow is consumer data via telematics to project based reporting.

Enterprise customers' willingness to pay for mileage and indicators has yet to be finalized. Part of this problem is conveying that specialized subsets of analytics actually cost more due to processing complexity, despite using less mileage and structured information as the foundation of these record sets. The telematics figures are an integral part of the corporation's intent to develop a full service dashboard.

Costs per vehicle are less than $100 for a subscription and $25 to $35 for smaller vehicle fleets. Pricing for aggregated records is currently based on a per mile basis at a cost of $0.05 per mile (historical facts). The foundation for historical statistics includes the make, model and year (on a per mile basis) for aggregated user and fleet trips. The service costs $0.07 per mile for (future figures) that is processed in real time. The processing for this service costs more and is marked up accordingly based on the cloud provider Amazon web services. The team wants to make their service and platform more frictionless and include transparency in pricing.

The organization plans on utilizing the financing from this round to hire new developers, including on the front-end primarily for web-resources, as the back-end is already largely completed. The business needs to be more specific regarding how the funding will actually be utilized. They are not presenting a solid case for their efficient use of capital thesis.

One concern with underwriting this start-up is the business is shifting its revenue strategy from large enterprise customers to smaller larger volume data driven sales. The fact that the start-up is still undecided about the exact pricing for their services at this stage of the game is a significant concern. Pivoting on its strategy at this juncture of development is disquieting for investors looking at this endeavor as a long-term investment. Venture financing shouldn't be used to determine appropriate pricing and to experiment with investors' money. The business should have a clearer picture of their revenue strategy. It would have been more advantageous for the start-up to establish their sales strategy and integrate this into their overall service offerings at year 1, rather than focus their sole efforts exclusively on the large Johnson Controls® and Ford® one-time enterprise contacts. This appears to be a strategic misstep. It would have been more advantageous to hire additional team members to focus on the recurring SAAS segment of the market earlier on. Instead, the endeavor is focusing on implementing this strategy after the enterprise contracts have dried up.

The battery diagnostic algorithm software is impressive, but may not be a significant enough source of monthly and annual revenue through licensing.

Conclusions

The business has developed a valuable data analytics platform and app technology. However, the value proposition would be stronger if there were able to develop models to improve traffic flow and actually make commuting a pleasant experience. Investment in this corporation would be conditional: they must make a reasonable effort to determine pricing first. After this strategy has been formally adopted an outlay in the $100 to $200K range would be warranted.

Cloud Asset Protection

A start-up I interviewed has been designing new technology focused on protecting assets in public cloud infrastructure. This company saves copies of measurements in the cloud. Backups are run during the night, meeting goals for encrypted protection. The start-up (which is at the inception stage) configures its protection to look for vulnerabilities in production systems. The technology constantly monitors existing environments and public backups by deploying its technology in an Amazon® architecture. Infrastructure costs are low because there is no hardware to be maintained, with the intended revenue stream being capacity based.

The firm is looking for funding in the form of convertible notes and the founder brings experience from his career working with Dell® and EMC®. The CEO developed technology that is used in EMC's® products. The technology from his current organization has no restrictions and no patents that prevent it from being deployed. The start-up has developed provisional patents, but doesn't have any formal patent protections in place.

The corporation acknowledged the difficulty of building partnerships, but has been in discussions with Accenture® regarding the deployment of its technology through the firm's service channels.

Conclusions

The start-up's value proposition is not clear and is far too early to gauge at this point in time. An investment in this endeavor is a solid pass. The firm has not made a clear case that it has the capability to solve many of the cybersecurity risks that are present in today's computing environments. As foreign attacks to U.S. assets become more and more prevalent it will become important to anticipate attacks before they occur and not retroactively attempt solve these problems.

Algorithm Based Rental and Roommate Platforms

This start-up[5] is using an algorithm to determine the most appropriate roommates. It costs $5 per month. The team has $200K committed, with a CTO and an additional business partner that targets partnerships and universities. However, the market is filled with competitors, including Roomi® that uses algorithms to match college students and young professionals with 4 to 6 month leases, completing all of the background checks and disclosures. Roomi's® program is nearing 2 million plus users and has raised $18 million in funding to date[6].

Significant problems with these suites are discrimination based on protected classes (race, religion, etc.), which have been highlighted by issues with bias on platforms, including Neighborly®. In the Neighborly® case African Americans experienced reporting based on their appearance in a community alone (not based on a specific infraction or incident of illegality).

---

[5]Confidential Discussion.5/14/18.
[6]Roomi® presentation 8/19/18.

One of the themes emerging in the startup community is the following: that bias in founders translates into harm into products. This is particularly true in housing, rental, real estate, fintech and vehicle on demand products. Founders should address bias head on and from the inception, bringing their teams together on a weekly basis to proactively remediate and address any discrimination from their platform and within their team.

Conclusions

Startups that are providing roommate or housing subdivision services are admirable. However, they are not truly innovative. To be radically disruptive in the marketplace they would be able to develop new housing and building materials that have not been seen in the sphere before (inventions that are incredibly inexpensive to manufacture, light, strong and durable). Building materials haven't changed substantially in years. Construction still uses wood or metal framing, sheet rock and concrete foundations.

Hardware Fabrication

This is an excellent priority chip manufacturing facility in Freemont, California. It has the ability to test, prototype and even develop schematics for chip design via outsourcing. Costs are reasonable and the team is able to troubleshoot problems with chip manufacturing and can provide sophisticated manufacturing remediation, including lead free boards. They do work for the military, government and the private sector, including international business partners in good standing with the U.S. (Israel). Manufacturing is done on-site and deliveries are completed using an unmarked white delivery van that is loaded with completed merchandise in a closed garage.

The facility has the ability to manufacture whiteboard and cardboard packaging (as needed) for consumer facing products. Costs are reasonable and chips for IoT devices can be assembled quickly and on-demand. Discounts are provided on a sliding scale and all products are manufactured in compliance with OSHA and other safety standards. Proprietary information is protected through a Mutual NDA that covers the manufacturing facility and the hardware start-up.

Each chip receives extensive trouble shooting for manufacturing defects prior to being shipped to the customer or retail outlet. This company also benefits from receiving the *Made in the USA* label, which is an excellent marketing strategy.

Conclusions

Many of these hardware start-ups have several of the key metrics that propel them forward from a stand-alone idea. They have one or more of the following: a minimum viable product, a solid team, user or client growth and/or actual revenue. However, factors that may create significant risk include: lack of regulatory approval, lack of a complete team, incomplete minimum viable product, unconfirmed growth or cap tables that are clouded by contingencies. It is for these reasons that many of these companies warrant revisiting in the technology and team in the future; and not making an investment immediately.

A better investment is to capitalize or take an equity stake in the hardware manufacturing facility. In the case of a manufacturing plant an underwriting strategy is recommended. There is a strong ROI in the facility itself. By taking an equity stake in the manufacturing facility an investor can in tern capitalize on deal flow leaving and entering the plant.

Artificial Intelligence (AI)

Artificial Intelligence (AI) has received a significant amount of funding. It does have useful applications in robotics and assembly line production; and for machines in dangerous environments: (highly toxic - noxious gases (sewers, fuel tanks); low visibility; radioactive; or extreme cold or heat).

However, its applications in law enforcement, education and employment are much more problematic. My attendance at a Google® DevFest in New Haven, Connecticut highlighted an examination of AI in the election. My question for the panelist was as follows: What about the use of AI in law enforcement in Wisconsin?[7] The question of fairness then became a part of this panel discussion.

---

[7]Adam Liptak. Sent to Prison by a Software Program's Secret Algorithms. Sidebar. The New York Times. May 1, 2017. https://www.nytimes.com/2017/05/01/us/politics/sent-to-prison-by-a-software-programs-secret-algorithms.html.

However, it is more than a question of fairness. It is an issue of ensuring *due process* for all civil and criminal defendants, as outlined in the 5th and 14th Amendments of the U.S. Constitution.

It is the court system and law enforcement that are engaging in discrimination and the denial of participants' *due process* – these are criminal actions in of themselves. There are rules for disqualifying and removing judges and law enforcement when they themselves are biased and/or in extreme cases are corrupt. For example, 28 U.S. Code § 455 - Disqualification of justice, judge, or magistrate judge must disqualify themselves in which that administrator's "impartiality might reasonably be questioned."[8] Further guidance for judicial conduct complaints is found under the Judicial Conduct and Disability Act of 1980 ("Act"), 28 U.S.C. §§ 351–364, and the Rules for Judicial-Conduct and Judicial-Disability Proceedings ("Rules")[9]. For example it is prohibited for a judge to retaliate "against complainants, witnesses, or others for their participation in this process;"[10] or "accepting bribes, gifts, or other personal favors related to the judicial office."[11]

---

[8] 28 U.S. Code § 455 - Disqualification of justice, judge, or magistrate judge. Cornell Law School. Legal Information Institute. https://www.law.cornell.edu/uscode/text/28/455.
[9] FAQs: Filing a Judicial Conduct or Disability Complaint Against a Federal Judge. US Courts. June 2016.
http://www.uscourts.gov/judges-judgeships/judicial-conduct-disability/faqs-filing-judicial-conduct-or-disability-complaint.
[10] *Id.*

Additionally, 28 U.S. Code Chapter 16 – Complaints Against Judges and Judicial Discipline[12], § 364 - Effect of felony conviction provides specific guidance for judges when a felony conviction has occurred.

In the infamous *Kids for Cash* scheme,

> Between 2003 and 2008, two Pennsylvania judges accepted millions of dollars in kickbacks from a private juvenile detention facility in exchange for sending children — girls and boys, some as young as 11 — to jail.[13]

This is a real example of why AI in the courts is highly prone to risk and exploitation, particularly for minority groups, which represent a disproportionate number (percentage) of the prison population.

---

[11]*Id.*
[12]*28 U.S. Code Chapter 16 – Complaints Against Judges and Judicial Discipline.* Cornell Law School. Legal Information Institute. https://www.law.cornell.edu/uscode/text/28/part-I/chapter-16.
[13]Abee Smith. Undue Process. *'Kids for Cash' and 'The Injustice System'.* The New York Times. March 29, 2013. https://www.nytimes.com/2013/03/31/books/review/kids-for-cash-and-the-injustice-system.html.

In effect judges and police are not immune from abuse, corruption or violence themselves. Accordingly there are rules and regulations regarding their discipline and removal. A system with closed AI data points opens up the possibility for discrimination against both plaintiffs and defendants. This type of system must receive a U.S. Supreme Court ruling on AI's application and its ability not to impede *procedural* and *substantive due process.*

Algorithms developed in secrecy do not guarantee a participant's right to *due process,* which is guaranteed under the U.S. Constitution. A defendant has the right to *procedural* and *substantive due process*; and a right to examine and contest the analytic points that make up a court-based algorithm decision.

The same is true for employment decision-making that relies on AI videos like a large New York multinational bank does for its pre-screening process. This extremely flawed, myopic and juvenile approach to hiring is highly likely to run into discrimination issues with the New York State Attorney General's Office, particularly because they ask the applicants if they are transgender and also about their sexual orientation. Not all exceptional candidates are extroverts; or demonstrate unique capabilities through a face scanning AI program. Many candidates have emotional intelligence and empathy characteristics that are not successfully demonstrated during this process; introverts, as an example. This extreme approach to HR needs to be removed immediately. Only George Orwell would be proud.

Additionally, using AI in employment (or similarly in education) screening tools (online systems) like Indeed® or online applications is a means of data collection, including misappropriating candidate's ideas and intellectual property. This is more than just unethical; it is *not* legal because it involves interstate commerce.

Many incubators and accelerators in the startup community use AI in the initial application process in the same manner. AI is used to screen out applicants, skim off the best ideas and pass them on to teams that already have executed contracts in place with those same incubators and/or accelerators.

As an example, one startup team that has solicited my oversight[14] submitted their proposal to an international incubator in Spain. They then advised me of the following: that a team with a similar name (with a contract already in place) used elements of their name and application to develop what they thought was a similar company. This is highly unscrupulous and illegal when United States Patent and Trademark Office (USPTO) protections are in place for the original applicants.

---

[14]Name withheld due to Non-Disclosure Agreement (NDA).

AI Skimming

This data extraction technique is what I would like to hereby define as *AI Skimming*. It is highly unethical, particularly because most companies are hiring from within and are only posting employment notices as a matter of U.S. employment law. They have no intention of hiring the online applicant, but their AI programs can skim keywords off of good applications that they can use to misappropriate intellectual property from websites, the applicant's data sets, including social media platforms like Twitter® and Facebook®.

The same is true of startup incubators and accelerators that use this *AI Skimming* technique to fraudulently strengthen their own portfolios. They have already selected companies from internal recommendations and are using the process to glean information from the competition.

Games, Photo and Video Sharing and Trivia

A large percentage of the highest level of traffic and profitability and network traffic falls under the games, photo and video sharing and trivia. Most of these apps are dopamine AI intensive technologies that utilize human insecurity and need for belonging to maximize their profitability.

Malware and other botnets are often hidden in these technologies, further exploiting vulnerabilities of the users technology and the overall network.

Conclusions

Some of the best uses of AI are in high-risk environments. Their deployment in employment, law enforcement, the judiciary and education require checks and balances. Those safeguards are not in place. AI is and will continue to be misused.

However, it is important to emphasize that the difficult and dangerous work that law enforcement agencies engage in not to be discounted or discredited.[15]

---

[15]My active donations include the following: CIA and FBI school scholarship and fallen officer funds (there are four of them).

Instead, it is recommended that the appropriate checks and balances be implemented whenever new technology is deployed in these environments.

Cybersecurity

Cybersecurity is an enormous issue with corporate and government espionage, hacking and theft. Cybersecurity has received significant VC funding over the last few years. The dollar amounts are as follows:

> 2017 was a record year for Cybersecurity funding, with total investments exceeding $3.6B. 2017[16].

This number can only be expected to rise as the number of retail operations, banks, corporations and government agencies are hacked. It is a major topic, even impacting national security.

There are a number of cybersecurity trends for 2018, including bots, botnets; phishing and macros; and ransomware; as based on a discussion with executives from a multi-national corporation and a leader in the area.[17]

---

[16]PwC/CB Insights. MoneyTree Report™ Q4. 2017.
[17]April 10, 2018 confidential discussion with multi-national corporation, unless otherwise noted.

Bots and Botnets

Bots and botnets target command and control CNC servers, based on a selection from 45,000 samples in the firm's analysis. A shutdown of this targeted disruption by bots by the corporation led to a 20 to 30% decrease in traffic and infiltration.

The question that remains is as follows: How quickly will cybersecurity companies and experts be able to reduce this traffic? This means reducing malicious botnet traffic from 30% to 60% while dropping the timeline for disrupting this infiltration from 3 months to 1 month (or less).

In a Distributed Denial of Service (DDOS) attack botnets try to remain undetected. For example one of the trends is to conduct bitcoin mining on an authorized server and/or server farm. This results in an increased electricity bill for the owners of the servers and computers.

Conclusions

These are serious issues.

On a residential scale this might include mining bitcoins from an elderly couple's computer without their knowledge on a fixed income. This could throw off their ability pay for food in the winter (or air conditioning in the summer) in order to cover the increased costs of their electricity. Botnet activity seeks to remain quiet, particularly for DDOS attacks that include unauthorized bitcoin mining.

On a national and international level bot and torrents can disrupt physical infrastructure (electricity, dams, et. al.), impact national security, food supply and banking services.

Cybersecurity is more than a part of a corporation's architecture. It is a core fundamental aspect of a firm's overall operation. In some instances it warrants having an entire division within the IT department that is exclusively focused on resolving these issues.

Macros and Phishing

Macros have made a return in the hacker community, due to their lower threshold of detectability. One of the principle areas for cyber attacks still remains phishing. These attacks have become more sophisticated, with emails remaining dormant until they have gone through a firewall, virus and malware software and then activating it an ½ hour later when users open it and/or it remains in the recipients email box. It involves timing in the sense that the malicious email first enters the system and then becomes malicious. The timing is about awareness.

Problematic Security Areas with Cloud Applications

This is particularly problematic with administrators and users moving to cloud based apps and software as a service (SAAS) because nearly 90% of all users aren't doing encryption across TLS (having deprecated SSL) with untrusted Wi-Fi networks.

It is essential that data protection be provided *at rest* and *in transit*, with insecurities that may exist within cloud applications. Encryption of this data *at rest* and *in transit* is the most important aspect of this process. Having a security policy in place is critical. It may require that the client clearly understand that their end user agreement requires that they encrypt portions of their data prior to transfer to the cloud. 87% of SAAS companies don't have these controls in place.

The CSA STAR Certification has been developed to rank companies based on their level of security (gold, silver, bonze).[18] Typically large corporations will advise their client users about what security and encryption they do and do not provide. It is a shared level of responsibility for most companies.

---

[18]About CSA STAR Certification. Cloud Security Alliance (CSA). https://cloudsecurityalliance.org/star/certification/#_overview.

Independent 3rd Party Audit

A 3rd party audit is essential in a facilities condition assessment, property evaluation and/or financial review or organization. A security systems audit is necessary for any organization. This cost can be more burdensome for smaller organizations but is absolutely essential.

Ransomware

Ransomware (holding data hostage) issues emerged in 2017 and 2018 with malicious code including *WannaCry*, *Petya/Not Petya* (which destroyed data) and *Bad Rabbit*. Ransomware is particularly problematic for hospitals, emergency response teams and law enforcement that become victims of this, because people's lives are at stake. Ransomware has also evolved to include lateral movements that include credential theft. The speed of these attacks has increased, becoming automated and significantly more destructive causing $100 millions of dollars in damages.

Not-for-profit hospitals and other smaller organizations are particularly vulnerable to these types of attacks. The Federal Bureau of Investigation (FBI) recommends that you contact them first; and that as an organization you do *not* pay extortion demands if at all possible (hospitals and other EMS organizations should examine this on a case-by-case basis).

It is imperative that organizations have a prevention and recovery strategy in place, creating destruction free backups and developing new host based firewalls that are isolated from the broader system. Organizations need to have systems that don't constantly have to be patched with updates. Advanced email protections need to be in place with a hardened host system that prevents malware from morphing to avoid AV sensors. Password protection of these systems is vital to ensure lack of unauthorized entry, and requires creating unique passwords.

In is also essential that more privileged admins are created, with just-in-time access. These types of systems ensure that they are only accessible when an organization wants to complete firewall and admin work; and create archived backups and restorations.

The faster that an organization completes service and software updates the better. They can isolate vulnerabilities more quickly in this manner; and make sure that threats are minimized. 3rd party SAAS apps also need to be audited to ensure that they are secure in their data collection in order to evolve how the organization's security approach is taken. Following these procedures a company can develop a rapid response and cyber resistance to malware; and reduce the mean time necessary to prevent and ensure data archival backups.

Many cybersecurity companies are developing multi-faceted tools that attack ransomware across infrastructure, hardware and software. Examining vendors to ensure that their online terms of service are aligned with a purchasing enterprise is critical.

An examination of the toxicity of isolated compromised data through a 3rd party audit is key. Furthermore determining if an organization has System and Organizational Controls (SOC) 1 (financial) and SOC 2 (non-financial security availability) controls in place is necessary. Security around API's is essential, particularly around data sharing and compatibility.

It is also crucial that a company have support in place for fraudulent calls. Calls can originate from multiple locations without being tied to the number and registered ID being shown on the phone's screen. This can lead to infiltration and extraction of sensitive data. Below is a photograph of a malicious spoofing phone call for aggressive data extraction.

Spoofing phone call from international call center threatening harassment and intimidation for data extraction.

However, email is still the primary mechanism for delivering ransomware and exploits. Machine algorithms have been developed to screen out most malicious software and to ensure that they have the right signatures in place. Companies need to be proactive, constantly evolve and develop new approaches to security. For example, one email click can quickly impact 60,000 to 70,000 machines. It also requires that an organization share non-proprietary issues and solutions with other virus and malware software companies. Included is a list of junk email that is part of a targeted DDOS attack. The intent is aggressive and the purpose is to harm the user with violence, threats and intimidation.

This message has no content.

● **Add Inches**   3:37 PM ›
Firmer and Naturally Bigger by 4 Inches... Try it Free.
This message has no content.

● **StopIRSDebt**   2:35 PM ›
IRS Announces NEW Fresh Start Programs for Taxp...
This message has no content.

● **Pre-Approval Notice**   12:45 PM ›
Up to $2,500 is Pre-Approved for Mlm292002
This message has no content.

● **CBD Drops**   10:31 AM ›
Potent Pain & Anxiety Relief - Free Trial for Mlm292...
This message has no content.

● **SexxyMommy153**   Yesterday ›
Have an affair with a local mom today
This message has no content.

● **Lyft Drivers Wanted**   Yesterday ›
$500 Bonus for Lyft Drivers after your first 100 rides!
$500 Bonus for Lyft Drivers after your first 100 rides!

⊜   Updated Just Now   ✎
    7 Unread

No Service 🛜 ☼   4:47 PM   ✳ 🔋
‹ Mailboxes   **Junk**   Edit

This message has no content.

● **Add Inches**   3:37 PM ›
Firmer and Naturally Bigger by 4 Inches... Try it Free.
This message has no content.

● **StopIRSDebt**   2:35 PM ›
IRS Announces NEW Fresh Start Programs for Taxp...
This message has no content.

● **Pre-Approval Notice**   12:45 PM ›
Up to $2,500 is Pre-Approved for Mlm292002
This message has no content.

● **CBD Drops**   10:31 AM ›
Potent Pain & Anxiety Relief - Free Trial for Mlm292...
This message has no content.

● **SexxyMommy153**   Yesterday ›
Have an affair with a local mom today
This message has no content.

● **Lyft Drivers Wanted**   Yesterday ›
$500 Bonus for Lyft Drivers after your first 100 rides!
$500 Bonus for Lyft Drivers after your first 100 rides!

⊜   Updated Just Now   ✎
    7 Unread

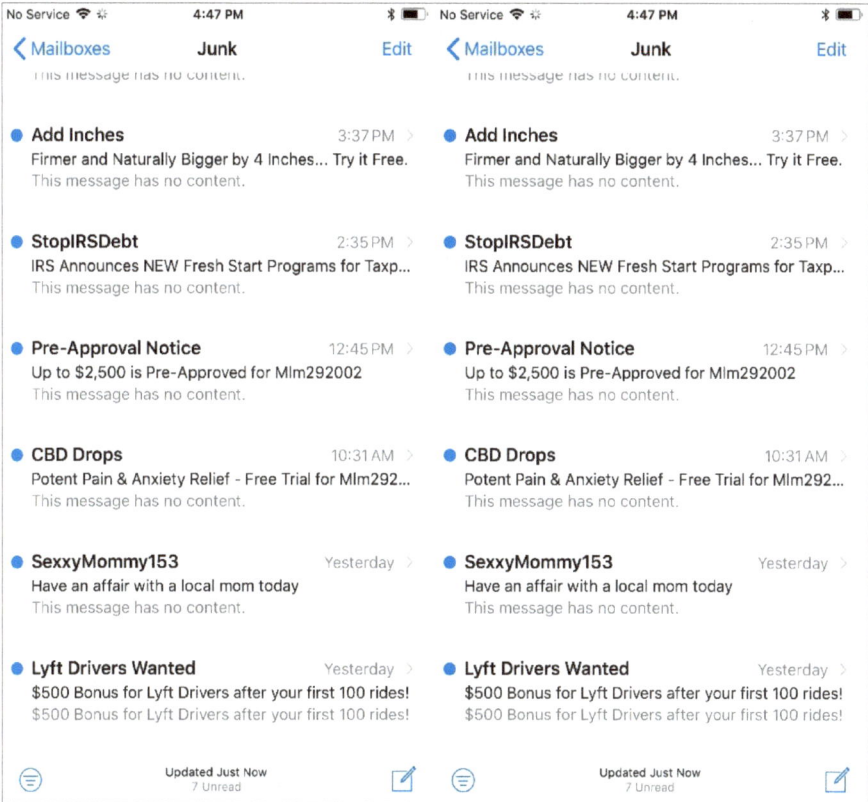

Malicious DDOS email attack executed against Yahoo!® account.

Recommendation: Contact Federal, State and Local Authorities

At this juncture the proper Federal, State and local authorities must be notified. The FBI, Federal Communications Commission (FCC) and Department of Justice (DOJ) are a few of the agencies that need to be contacted. Take careful notes, photograph (the phone's caller ID information and take a screenshot of the emails) and document the dates of the threatening, harassing and intimidating calls and emails. Forward the information in stages to law enforcement and regulatory agencies. Try not to engage with the harassing parties (even if there are threatening violence). Remain calm.

For elderly people that are rushing to respond to phone communications this creates a particularly dangerous situation if they fall and/or are disabled in the process. On March 6 and 7, 2018 and again on April 6, 2018 a colleague received threatening and terroristic demands from Indian call centers (and other locations throughout the United States based on area code) immediately after website registration contact information updates were made.[19] The calls originated even after three phone numbers were changed to protect these individuals (i.e. new numbers were issued). This colleague received multiple requests from automated messages purporting to be from law enforcement (and/or otherwise) indicating that they would be locked up.

The messages originated (and continue to initiate from) from domestic and international numbers meeting the test of 47 U.S. Code § 223.[20] The statute is outlined as follows: *47 U.S. Code § 223 - Obscene or harassing telephone calls in the District of Columbia or in interstate or foreign communications*[21], prohibits these types of threatening and harassing phone calls.

---

[19]Actual Federal complaint documented by colleague and filed with FBI, DOJ and FCC and four (4) multi-national corporations. Ongoing legal complaint. Names withheld.
[20]Cornell University. Legal Information Institute. 47 U.S. Code § 223.
https://www.law.cornell.edu/uscode/text/47/223
[21]*Id.*

Conclusions

Threatening phone calls are a common pattern and the attacks become more and more aggressive over time, with spoofing numbers and private VPNs tied to digital phone numbers used to start the violent attacks. The verbal assaults (an international and interstate felony) start with automated dialing and then a physical (actual) person is added to the line, yelling and screaming further spoken blitzes. The assailants attempt to extort money threatening physical restraint with fictional law enforcement. They robo-call their targets across multiple masked numbers until they are successful and receive a wire transfer. The underlining infrastructure for these encroachments is the telephone and technology companies manufacturing these devices. The responsibility falls under their legal purview.

Venture Capital Tiers

Venture capital firms provide funding (a form of private equity) for start-up companies of varying sizes. These firms obtain capital in order to complete their activities from individual and institutional investors;[22] and tend to invest by specialized stages.[23]

There are approximately four types of venture capitalists (and/or private equity firms). Dow Jones & Company, Inc.® defines the first as seed-stage firms that complete early term sheets as VC's who:

> Tend to provide a few hundred thousand dollars, and perhaps some office space, to an entrepreneur who needs to flesh out a business plan.[24]

---

[22]David M. Toll, Managing Editor, Dow Jones Company. Private Equity Primer. *Galante's VC & Private Equity Directory.* Dow Jones & Company, Inc. 2005.
[23]*Id.*
[24]*Id.*

Many of these companies are unsuccessful. Ultimately many of the target audiences for these enterprises are too niche and do not have the infrastructure or supply chain mechanisms in place to support their operations over the long-term. Burn rate is particularly applicable here because companies in many cases pay for expensive real estate, programmers and advertising. This overhead eats into their equity (an/or equity/debt combination) financing and quickly reduces their profitability in the first few years of operation. Businesses, in some cases seek bridge financing, if they have obtained a market fit. This funding is essential to tide these firms over until they secure an angel and/or Series A-Round. Rounds are typical denoted by letters of the alphabet and are used until one of the following occurs: business closure, aqui-hire (buying a company for the founder and/or staff), later B, C or D Round, Merger & Acquisition (M&A) occurs; or an Initial Public Offering (IPO) on one of the exchanges.

The next stage includes early-stage investors. In some instances this may involve angel investors or backing to very-early stage businesses.[25] This group of underwriters provides equity financing for the following:

Companies at a point where they have a completed business plan, at least part of a management team in place, and perhaps a working prototype.[26]

The requirement for a just a working prototype has shifted more to requiring:

1. A working minimum viable product (MVP);
2. A team;
3. Monthly Recurring Revenue (MRR) up to $25K; and
4. Acceptable growth metrics (user and/or client).

Investors do underwire companies that do not meet these criteria. However, it has become harder and harder to secure an investment without doing so.

Dow Jones® characterizes the next group as late stage-round investors. This group:

---

[25]*Id.*
[26]*Id.*

Provides a second or third-round of financing, often of $10 million or more, that funds production, sales and marketing, and carries the company into the revenue-producing stage.[27]

In Series A rounds and higher MRR shifts to approximately $100K and higher, with corresponding higher user and client growth metrics.

Finally, at the Mezzanine, or pre-IPO stage, investors provide a final round of financing that transports the corporation to an IPO.[28] Venture capital has served (in many instances) as a platform for an Initial Public Offering (IPO). For example, Sequoia Capital® is a California based venture firm that capitalized YouTube® that was acquired by Google® in 2006 for $1.65 billion. Sequoia Capital® received substantial returns from this speculation.

---

[27]*Id.*
[28]*Id.*

Sequoia®, which is among the most successful venture firms in Silicon Valley, invested a total of $11.5 million in YouTube® from November 2005 to April 2006. It may be walking away with more than 43 times that amount. Its stake in YouTube® has been estimated at roughly 30 percent, which would give it a value of $495 million[29].

Sequoia Capital® has subsequently financed Dropbox® which executed an IPO and Zappos®, which was acquired by Amazon®. Sequoia Capital Global Growth Fund II, L.P. filed a Form D with the SEC for an additional $1,999,5 billion on June 1, 2017. [30]

[29]Miguel Helft and Matt Richtel. *Venture Capital Firm Shares a YouTube Jackpot*. The New York Times. October 10, 2006.
[30]Form D. Securities and Exchange Commission (SEC). Sequoia Capital Global Growth Fund II, L.P. June 1, 2017.

2005-2009 Updates

Additionally, a 2007 *Global Venture Capital Insights Report* produced by Ernest and Young® noted "the pool of private venture-backed companies in the United States stood at 5,476 companies with a cumulative U.S. $131 billion invested as of January 1, 2007".[31] As of 2005, there were over 900 venture capital firms in the United States.[32]

2012-2017 Developments

The breakdown for U.S annual VC funding was as follows: $32.6 billion (2012), $36.1 billion (2013), $59.4 billion (2014), $76.8 billion (2015) and $61.4 billion (2016).[33] In 2017 this investment dollar amount reached $71.9 billion.[34]

---

[31] John D. Yonge. The Global Pool of Private Venture-Backed Companies Approaches 10,000. Acceleration: The Global Venture Capital Insights Report 2007. Ernst & Young. 2007.
[32] *Id.*
[33] PwC/CB Insights. MoneyTree Report™ Q4. 2017.
[34] *Id.*

VC's Critical Role in Stimulating Growth

Venture capital firms play an essential role in stimulating growth in the U.S. economy: The growth of specialized investment firms was aided by the development of the federal Small Business Investment Company (SBIC) program in 1958.[35] The SBIC Program is a financial assistance program made possible by the U.S. Small Business Administration (SBA).[36] This program consists of SBICs that "are federally licensed venture capital firms that can borrow money with a government guarantee of repayment".[37]

---

[35]*Id.*
[36]Entrepreneurs Seeking Financing. The SBIC Program: Seeking SBIC Financing for your Small Business. U.S. Small Business Administration.
http://www.sba.gov/aboutsba/sbaprograms/inv/esf/inv_sbic_financing.html.
[37]*Id.*

As of 2009 there are over 400 licensed SBIC's in operation.[38] In 2011, the FY for which the U.S. Small Business Administration (SBA) provided numbers was 299, with a total of $8.2 billion invested.[39] The SBA has not provided updated numbers for later years, including 2018.

Ascent Venture Partners® is an example of a licensed SBIC business. Ascent Venture Partners® "has been investing in early stage, emerging technology companies since 1985".[40] The VC fund has backed over 90 companies and has managed four venture funds with total outlays of over $400M.[41]

As of 2017 this number had jumped to $500 million under management.[42] It is an example of a company that initially fulfilled the SBA's programmatic mission.

---

[38]*Id.*

[39]About SBA Fact Sheet. 2018. https://www.sba.gov/sbic/general-information/faqs#2.

[40]Overview Fact Sheet. Ascent Venture Partners. 2007. http://www.ascentvp.com/.

[41]*Id.*

[42]Chris Witkowsky. Harbert Growth, Ascent Venture, Osage Venture lead $11 million funding in Sidecar. The PE Hub Network. May 11, 2017.

The SBA's guidelines define companies that are eligible for support based on their size. The SBA's SBIC program funds companies like Ascent Venture Partners® because they are small. The SBIC's definition of a small company is one whose "net worth is $18 million or less and its average after tax net income for the prior two years does not exceed $6.0 million".[43] SBIC's typically select a particular area of expertise and a particular stage of funding.[44] Ascent Venture Partners® focus is on early stage and emerging technology companies as noted above.

Magnet Capital Partners® concentrates on Mezzanine capital, which is defined generally as follows,

> Mezzanine capital, also known as 'Subordinated Debt' is generally considered a hybrid of debt and equity. It is a loan that is often unsecured (or minimally secured) and is at greater risk than 'senior debt' that is customarily provided by banks, lease companies, asset lenders and other similar sources.[45]

---

[43]*Id.*
[44]*Id.*
[45]About Us. Magnet Capital. 2009.
http://www.magnetcapital.com/fund.html.

Magnate Capital Partner® manages a Mezzanine Fund of over $20 million. They are searching for investment and lending opportunities with existing companies that are seeking capital for growth, expansion and buyout".[46] This is the last stage of venture capital investment prior to an Initial Public Offering (IPO). Magnet Capital Partners® is targeted on high-tech, low-low tech and no-tech manufacturing, distribution and wholesale, and service industries.[47]

More than a thousand new independent private venture capital companies that don't rely on government support have supplanted the SBIC's, which emerged in the 1950's.[48]

---

[46]*Id.*
[47]Candidate Profile. Magnet Capital. 2009. http://www.magnetcapital.com/profile.html.
[48]About SBA Fact Sheet. 2018. https://www.sba.gov/sbic.

Conclusions

The SBA and other government agencies should continue to provide acceleration zones, working spaces, loans and other instruments to promote a start-ups, M&A and IPO activity. This is particularly important with the number of publicly listed companies on NYSE, NASDAQ and other exchanges being reduced by several thousand from 8,000 +.

The smaller number of companies negatively impacts the positive multiplier effects that a larger number of companies provide in the marketplace in terms of competition and growth. It's simple: making an investment in the business infrastructure of a community is essential in the same way that corporate R&D provides new innovation.

Sources and Uses of U.S. Venture Capital Funds

Venture capital funds are provided by a variety of sources. Some of the principle sources are pension funds, foundations and endowments, corporations, families, individuals, banks and insurance companies.[49] Since 1990 more than 40 percent of the capital has come from pension funds followed by endowments and foundations "with an average of about 20 percent of the funds each year".[50]

---

[49]*Id.*
[50]*Id.*

2007 Cleantech Allocation

As of 2007, in the United States the allocation of capital invested among primary industries is "about 60% in information technology (IT), 30% in life sciences, and 10% in business consumer/retail (BCR)".[51] Additionally, venture capital funds tend to have a distribution of funds that is based on company life cycle. As of 2007 this distribution was as follows: "30-40% in early stage rounds, 20-25% in second rounds and 35-40% in later rounds".[52] In 2007 a total of 56 venture-backed IPOs raised $3.7 billion.[53] Venture capitalists and their associated firms have focused their fiscal allocations (uses of funds) in several areas.

---

[51]Gil Forer, Ernest & Young and Dr. Martin Haemmig, CeTIM. Acceleration: Global Venture Capital Insights Report 2007. Ernest & Young. 2007.
[52]*Id.*
[53]*Id.*

These areas include clean technology (or Cleantech).[54] The U.S. invested a large percentage of the $1.8 billion that was raised in "140 financing rounds in 2006 in China, Europe, Israel and the United States" for this type of venture capital funding.[55] Clean technologies are a new phenomenon and capital deployments are related to the fluctuations in oil prices and the technologies economies of scale; and are focused in the area of "energy and water".[56] In the clean technologies H2Gen Innovations is a U.S. based firm that is financed by venture capital that specializes in fuel cell and energy production.[57] H2Gen Innovations, Inc. provides "reliable, cost effective plants for hydrogen generation and gas purification".[58] In 2007, according to the *Cleantech comes of age* report by Pricewaterhouse Coopers®, the U.S. venture capitalists "poured $2.2 billion into US Cleantech companies".[59]

---

[54]*Id.*
[55]*Id.*
[56]*Id.*
[57]*Id.*
[58]Home. H2Gen. 2009. http://www.h2gen.com/.
[59]Cleantech comes of age: Findings from the MoneyTree Report – A Quarterly Survey Produced by PricewaterhouseCoopers® and the National Venture Capital Association® based on data provided by Thomas Reuters®. PricewatershouseCoopers®. 2008.

Cleantech deals in 2017, if they are funded at all have become larger isolated deals, SolarCity's® acquisition by Tesla®, for example.

2017-2018 Cleantech Allocation

VC Cleantech investments have shifted in the 2017 - 2018 period as follows,

> Between 2011 and 2016, VC cleantech investment declined by nearly 30 percent, from $7.5 billion to $5.24 billion[60].

By comparison, Internet companies received $6.5 billion in 2017; and software $1.6 billion.[61] In Q4 2017 U.S. VC underwritten companies raised $18.7 billion.[62] This reflects an increase over a ten-year period and a concurrent decrease in cleantech over the same general period. Despite this overall decrease in cleantech, the renewable energy and solar sector continues to include new deals & technology.

---

[60]Devashree Saha and Mark Muro, Cleantech venture capital: Continued declines and narrow geography limit prospects. Brookings. May 16, 2017. https://www.brookings.edu/research/cleantech-venture-capital-continued-declines-and-narrow-geography-limit-prospects/.
[61]PwC/CB Insights. MoneyTree Report™ Q4. 2017.
[62]*Id.*

Solar Hardware

One of the trends in venture funding includes companies that are developing hardware to increase the efficiency of solar installation times. A business in this space has developed a technology that can improve this efficiency by 50% - a 2x increase.[63] This firm has a second product, which has been shipped to Thailand, Germany, and Australia; and is waiting for U.S. certification. The corporation's market breakdown is as follows: 52% industrial, 30% residential and 18% commercial. The manufacturing, shipping and other freight forwarding activities are outsourced for one of the firm's products. The other device is factory-made in Connecticut and is only 10% more expensive to manufacture in the U.S. with ISO 9001 certification.

---

[63]Confidential deal screening. 7/12/18.

The products are composite fixtures that are a combination of metal and plastic. This composite material is a concern for sunlight durability. In California and Hawai'i extreme temperatures and sunlight cause plastic to degrade rapidly. This company's product is rated for 18,000 hours of sunlight wear and durability (or approximately 9.3 years) with a 10% drop in strength over time. One concern is a replacement and maintenance cycle, particular due to the 10MW of solar installations that the startup has completed as a part of its sales cycle.

The start-up engages in direct sales to installers, including those that are commercial and industrial, including Walmart® and factories. It has a pending 125 MW utility scale installation in Asia and a projected $10 million in sales. The manufacturing margins are 40%, which are some of the highest in the industry. The enterprise is raising $250K (and has firm commitments on $250K), based on a $6 million pre-money valuation.

Solar Panel Innovations – Integrated Shingles

Another start-up has it own integrated solar panel shingles, where the panels are embedded into the roof.[64] The difference between Tesla® and this firm is that their product is actively sold online. Tesla® has a competing product with 8 MW demand and $40 million dollars in reservations.

It has two products: a shingle and a tile, each of which are in production. Both products are very durable in extreme conditions, including hail.

This solar panel manufacturer has $1 million in revenues. They have also received a $500K loan at 1.5% as a part of a NYS grant award. The start-up is seeking $2 million in growth capital.

One principal concern for this firm is that the Federal tax credits expire in 2021. This means that for many residential and commercial customers the incentives to purchase the technology will shift the demand curve for these products.

---

[64]Confidential deal screening. 7/12/18.

The firm sells directly to dealers and other integrators. The supply chain is broken down as follows. The panels are manufactured in Poughkeepsie, New York due to the newly implemented tariffs. The endeavor has one patent and two that are in the process of being completed. The corporation's products sell for a 10 to 20% premium. This is a reasonable margin for this type of manufacturing output and material.

Solar Canopies Combined with Charging Stations

My due diligence on another start-up at a venture presentation included an entrepreneurial endeavor selling two electric energy-charging stations combined with a solar canopy. For a typical office complex this includes 4 parking spaces and will accommodate most of the major brands, including Tesla® and BMW®.

The charging canopies standout from the typical competing units because they can fully charge a vehicle in 10 minutes. This differs from the typical municipal unit that takes several hours.

The stations provide 160 KWh of electricity through a partnership with BrightField. They compete directly with Tesla's®. The firms' stations work with Tesla® cars and other major manufacturers, costing $18 to $20 to fill up a tank, compared with $0.19 per KWh to charge at home. Profit margins are 200%.

The firm's niche is the 200-mile user who is seeking this range of electric car accessibility. The organization is developing a network with an app to find stations for users, clients and customers. By 2022 the corporation is seeking to capture 8% of the market with 9MW of capacity.

The business is a capital expenditure (CAPX) for investing companies. The cost is $365K for a canopy and two charging locations, including 4 parking spaces. The company can provide service for up to 15 cars per day per station. The solar panels typically carry the cost of financing. The start-up has indicated the following: that their profit margins will not be significantly impacted by the phasing out of the solar tax credits in 2021; and that their business is a cash in cash out corporation.

The start-up is seeking capital to fund 13 centers. This includes developing 5 MW of wind and solar.

Solar and Renewable in 2018 - 2019

Despite the downturn in Cleantech as a venture capital focus area it can be anticipated that investors will continue to finance deals in the renewable energy space and associated infrastructure. This can be expected despite the loss of Federal tax credits in 2021, which will shift the demand curve for these products back. This loss in tax credits will be offset (in part) by economies of scale, which will push renewable energy costs down further, making them more cost effective for residential, commercial and industrial installations.

Biotechnology in 2008

Biotechnology is a third area that has experienced rapid growth in the venture capital sector. Gil Forer, Ernest & Young® and Dr. Martin Haemming, CeTIM's report notes that globally (including the U.S.):

> Public and private equity investors invested handsomely in the sector in 2006, driving total capital raised to US$28 billion – a massive 42% increase over 2005, and second only to 2000, when the industry was at the height of the genomics bubble.[65]

In 2008, growth in the biotechnology sector stalled with a decrease of investments of $100 million from $1.7 billion (2007) to $1.6 billion (2008).

Biotechnology in 2017 and Beyond

The biotechnology sector remains one of the principle areas that utilize venture capital. Genomics (including DNA analysis) is one biotechnology and science focus area that has received funding. In 2017, Genomics companies received record funding in 2017 at $2.5B[66]

---

[65]*Id.*
[66]*Id.*

An investment presentation that I went to in New York City included a company that claims to increase the percentage of DNA that the presenting endeavor could provide human analysis with accuracy of up to 100%.[67] Typically in science there is some threshold that is nearly 100%, but never reaches it. In medicine, for example, it is not always possible to immunize everyone 100% with a vaccine. They also made the claim that Food & Drug Administration (FDA) approval was not required. Further due diligence is required for the investment community on deals like this.

For example, CV Therapeutics is a U.S. based biotechnology firm that is financed by Versant Ventures®, which specializes in healthcare and biotechnology venture capital.[68] CV Therapeutics is a biopharmaceutical corporation "focused on the discovery, development and commercialization of new small molecule drugs for the treatment of cardiovascular diseases".[69]

---

[67]Corporation's name and event withheld. 2018.
[68]Previous Investments. Versant Ventures: Healthcare and Biotech Venture Capital. Versant Venture Management LLC. 2007. http://www.versantventures.com/legacy_companies.html#.
[69]About. CV Therapeutics, Inc. 2009. http://www.cvt.com/.

Biotechnology has continued to receive venture capital attention, but attention has concurrently shifted to PAAS, SAAS, AI, VR, Agtech and Fintech in 2017 and 2018.

Software as a Service (SAAS)

Software companies continue to be a target of venture capitalists. They take very little capital to start and there is an abundance of programmers and developers that can code across multiple languages. This has flooded the marketplace due to the reduction of the cost of entry but has also created a glut of mobile apps and platforms, requiring aggressive advertising and extensive capital to stand out from the competition. There was approximately $850 million raised in 2006 for SAAS and PAAS companies and the majority of the growth was in the U.S. and Europe.[70]

---

[70]*Id.*

Major Players in the Industry (Major Corporations or Decision Makers)

The major players in the U.S. venture capital industry are the U.S. Small Business Administration (SBA)'s Small Business Investment Company (SBIC) program and its approximately 299 (down from 400) associated organizations. Other industry players include the multitude of private venture capital firms that has expanded exponentially. As noted previously there were 5,476 firms "with a cumulative US$131 billion invested as of January 1, 2007".[71] This represents a "decrease of 7% in the number of businesses and 8% in the cumulative capital invested in them since 2003".[72]

---

[71] *Id.*
[72] *Id.*

2017-2018 Allocation

In 2018 private fundraising (which includes Venture Capital as a subset) has provided approximately $2.4 trillion in capital for corporations.[73] This number reflects a gradual shift from Initial Public Offerings (IPOs) to private offerings, in addition to venture capital equity offerings from the larger firms, including, but not limited to, Sequoia®.

---

[73]Jean Eaglesham and Coulter Jones. *The Fuel Powering Corporate America: $2.4 Trillion in Private Fundraising.* The Wall Street Journal. April 2, 2018.

Irish Tax Law

Ireland continues to be a noteworthy source of U.S. startups seeking to offset corporate taxes through their substantial corporate tax advantage of 12.5%. Despite the tax changes in the U.S. there are still over 700 U.S. businesses with over $387 billion in total asset stock of the U.S. in Ireland.[74] Ireland represents 1% of the E.U economy but has attracted 12.1% of all foreign direct investment (FDI).[75]

As a consequence Ireland continues to represent a significant opportunity for start-ups and established technology businesses, not just Apple® and Google®. Ireland provides a competitive advantage for emerging tech firms and it has the administrative framework in place to support this.

---

[74]Private tax presentation. 6/26/18.
[75]*Id.*

Conclusions

Startups, entrepreneurs and established firms should all take advantage of Irish and other offshore tax technologies. To the extent that entrepreneurial endeavors can utilize the benefits of the revisions in U.S. tax law to repatriate funds they should also do so. Regardless for emerging companies it is worth having multiple accountants and firms to handle tax issues as they emerge.

Hedge Fund Analysis

2018 – 2019 Trends

Building a hedge fund is a challenging proposition. It is primarily about relationships, with a clear business proposition of how capital is being put to work. In many instances a family office will not elect to invest in a manager's first fund, waiting to understand the performance of funds II and III. Managers should carefully prepare their due diligence to target the most appropriate type of asset class and fund (i.e. including, but not limited to Taft-Hartley, Muni, University and Pension system Funds, et. al.). Every family-office is different, with European offices being older and particularly more risk-averse. Funds at the extremely small end of the family office spectrum range from $5M to $500M.

Institutional hedge funds can include assets of several trillion dollars and the tests of a good manager are the same:

1) What is their track record?
2) How have their funds performed over time?

3) Do they have an intimate knowledge of their investment (site visits – real estate, stadiums and apartment complexes)?

The issues that hedge funds and their managers' face are similar to any start-up and/or entrepreneurial endeavor. Funds need to obtain insurance and prepare their own internal due diligence. Lock-ups are common and as a consequence the selection of assets to invest in is particularly important as their performance will be judged overt time (i.e. Vintage funds).

It is essential to implement best practices to efficiently and proactively source capital. Several fund managers recommend using CRM for their scheduling and follow-up practices.

In preparing a fund strategy it is important to complete the due diligence. The two strategic questions for deployment of capital are as follows:

1) Is the strategy defensible or not defensive?
2) Is it an efficient use of capital?

These questions should constantly be examined and re-analyzed. Funds should reallocate their capital deployments in all instances where the use of capital is not efficient and/or defensible.

Examples: Inefficient Use of Capital and Defensive Pricing Strategies

Several new start-ups that I have been in discussions with[76] have developed a new technology. They provide an excellent example of the inefficient use of capital and a defensive rather than defensible strategy. In one instance it is their only product in their queue. In another examined case the corporation had not finalized the direct to consumer pricing strategy over a two-year period.

---

[76]First discussion with this company commenced on September 17, 2018.

After reviewing this entrepreneurial endeavor[77], it has become clear that their fiscal projections are unrealistic and lack an understanding of basic economic principles. The business included funding (as a part of their proposal for venture allocation) for three (3) years of Research & Development (R&D). However, during this period the company only represented one consistent product pipeline.

The expectations of private equity firms, hedge funds and other publicly traded investment firms are clear: R&D must produce new product pipelines over time. In this case the company had not included any other new products during the three (3) year period.

This is a fundamental flaw in financial modeling and would have instantly resulted in their proposal being sidelined. Organizations that are publicly traded already receive heavy criticism of their R&D spending. To highlight unproductive R&D at the every early stage sends the wrong signal to investors and torpedoes their chances for success.

---

[77]November 5, 2018 discussion.

Unfortunately it is similar to other corporations that are fundraising without a clear valuation strategy. Another early stage business (referenced earlier) obtained funding for two large corporate projects with Ford® and Johnson Controls®. During the financing presentation the startup acknowledged that it intended to establish pricing based on testing their cost strategies directly in the marketplace. However, the start-up should have begun the process of establishing appraising from the onset of its founding, using some of the proceeds from their initial contracts to determine the appropriate product estimating for their current launch.

My confidence in their ability to quickly determine service appraising in the marketplace was not there. Their lack of willingness to discuss their deployment strategy with me was a significant concern. At this juncture without corrective action investment is not an option. The enterprise is not demonstrating that it can use capital efficiently. Their R&D strategy highlighted their major weakness: an inability to think strategically and to demonstrate that they can use investors' money wisely.

A Comparison: Major Venture Capital (VC) Industry Trends Over Time

Investors are constantly seeking new and innovative methods of deploying capital. This is manifested in their attempts to categorize disruption by segment, size and stage. Categories that once attracted significant capital deployment, including Cleantech have shifted into sustainability in housing, real estate and transportation. Investors' appetite for risk continues and many categories that didn't exist ten years ago, or were not invested heavily in, have emerged: smart-car logistics, apparel, restaurants, wearable devices for health and fitness; and in a multitude of other sectors. Many private equity and venture capital speculations are illiquid, meaning that investors are unable to easily move in and out of funds. Returns are based on long-term deployments of capital and in some cases a merger and acquisition.

2007-2009 Trends

Venture capital investment trends in the U.S. include, but are not limited to, funding of cleantech, software/Web 2.0 and biotechnology. It is important to note that cleantech and biotechnology have not completely supplanted venture capital assets in software/Web 2.0. Software, according to a 2008 fourth quarter MoneyTree Report, still exceeded other investment categories like biotechnology, industrial/energy and medical devices and equipment.[78] In 2008 venture capital funds in software were $1,021 billion.[79] Web 2.0 start-ups received substantial venture capital funding for start-ups and companies in other stages of their development. For example:

> Venture capitalists put $1.34 billion into 178 deals in 2007, an 88 percent jump over 2006. But once you strip out the $300 million that Facebook raised from Microsoft and others, the numbers don't look as bullish.[80]

---

[78]*Investments by Industry. Q4 2008.* MoneyTree Report. PricewaterhouseCoopers. 2009.
https://www.pwcmoneytree.com/MTPublic/ns/nav.jsp?page=industry.
[79]*Id.*

Web 2.0 is defined as the web as a platform.[81]

This is now (in the present) typically defined as PAAS and SAAS. It is an area that has been invested heavily in after the speculative market of the dot.com financing experienced a sharp decline in 2001.[82] Platform as a service (PAAS) and Software as a service (SAAS) have been an area of continued capital deployment, but may experience a downturn that coincides with current economic and market conditions.

---

[80]Martin LaMonica. *Is Venture capital's love affair with Web 2.0 over?* News Blog-CNET News. March 18, 2008. http://news.cnet.com/8301-10784_3-9896632-7.html.
[81]Tim O'Reilly. *O'Reilly Network: What Is Web 2.0?* O'Reilly. September 30, 2005.
http://www.oreillynet.com/pub/a/oreilly/tim/news/2005/09/30/what-is-web-20.html.
[82]*Id.*

Cleantech & Biotechnology: Categorical Leaders a Decade Ago

Cleantech and biotechnology previously were categorical leaders of funds for venture capitalists. In the fourth quarter of 2008 biotechnology funds were $1,016 billion.[83] They have grown in relationship to venture capital speculations in software. Cleantech was one of the fastest growing venture capital sectors from 2002 to 2007. In 2002, the sector grew from $263 million of funds to over $2.188 billion in 2007.[84] Cleantech has the following sub-sector trends: (1) Solar; (2) Wind; (3) Power supplies (batteries, fuel cells, et. al.); (4) biofuels; and (5) Pollution, recycling and watertech.[85] Solar energy as a subsector "attracted the largest share of cleantech investment, with nearly $600 million committed in 39 deals".[86] In 2008, cleantech "experienced significant growth in 2008 with $4.1 billion invested in 277 deals".[87]

---

[83]*Id.*
[84]*Id.*
[85]*Id.*
[86]*Id.*
[87]*Id.*

The National Venture Capital Association (NVCA) annual survey noted that the industry was likely to experience a downturn in venture capital funding in 2009 as the "country's economy, the capital markets, and the venture industry as the global financial crisis takes its toll on the entrepreneurial ecosystem".[88] Additional evidence of this is provided by PricewaterhouseCoopers in their Cleantech Index (an index tracking publicly listed cleantech startups, compiled by the Cleantech Group LLC) which "rose by 42.4% for the year, compared to the NASDAQ Composite's 9.8% gain and the S&P 500's 3.5%".[89] This sector declined in relationship to the broader markets in the first two quarters of 2008.[90]

---

[88]Laura Cruz and Emily Mendell. *Venture Capitalists Predict a Difficult 2009. NVCA Annual Survey Forecasts Challenges for Venture Industry in the Coming Year*. National Venture Capital Association (NVCA). Washington, D.C. December 17, 2008.

[89]Figure 7. Publicly traded cleantech companies make big run in 2007, stumble in early 2008. Cleantech comes of age: Findings from the MoneyTree Report – A Quarterly Survey Produced by PricewaterhouseCoopers and the National Venture Capital Association based on data provided by Thomas Reuters. PricewatershouseCoopers. 2008.

[90]*Id.*

Cleantech venture capitalist money deployments then declined from 2008 through 2016. There was some decline in investors overall willingness to invest in this sector during this time. However, economies of scale in the renewables sector continue to drive investment even with shifting interest.

Cleantech Trends Now

In 2017, as noted previously between fiscal years 2011 and 2016, VC cleantech speculation declined by nearly 30 percent, from $7.5 billion to $5.24 billion.[91] This is still a very high level of investment on the part of the VC community and is not expected to decline.

Renewable energy and new energy sources can be expected to emerge as new areas of interest for venture capitalists, including net zero housing. This is defined as follows: housing that is completely self-reliant (or producing energy to offset any use) by being completely energy efficient in the summer and winter, including solar panels and solar water heating and exceptional insulation.

---

[91]Devashree Saha and Mark Muro, Cleantech venture capital: Continued declines and narrow geography limit prospects. Brookings. May 16, 2017. https://www.brookings.edu/research/cleantech-venture-capital-continued-declines-and-narrow-geography-limit-prospects/.

New Area: Efficient Sustainable Net-Zero Affordable Housing

Efficient Sustainable Net-Zero Affordable Housing will emerge as a new stand-alone asset category in 2018, 2019 and beyond. I am defining this category as its own breakout-underwriting group. The need for this type of housing is particularly acute in California, Louisiana, Texas, Puerto Rico, New Jersey and other communities that have been impacted by significant climate events. The fires in California have destroyed 14,000 homes and caused $3 billion in damages in 2017.[92] 2018 is gearing up for an exceptional summer in California with nearly 90,000 acres burned over a seven-day period in June and July; and the even more exceptional and tragic Paradise, California fire thereafter.

---

[92]Richard Wilton and Shelby Grad. *Losses from Northern California Wildfires top $3 billion; 14,000 homes destroyed or damaged.* LA Times. October 31, 2017.
http://www.latimes.com/local/california/la-me-california-wildfire-insurance-claims-20171031-story.html.

The flooding and hurricane damages in the southern United States and Puerto Rico caused billions of dollars in damages. There is a need for housing to replace all of these destroyed structures. This housing void will be filled (in part) by this new deal class, which combines fire and flood mitigation with net-zero energy efficiency.

Conclusions and Technology Limitations

The following was showcased as a guest at an investment presentation: a proposal for equity shares (a percentage stake) in a net zero housing corporation.[93] Affordable net zero housing typically includes computer numerical control (CNC) cut windows, doors and roofs to improve the energy efficiency of the structure. This is a solid investment and worth pursuing.

One limitation of this type of super efficient design involves the use of pressboard cabinets, furniture and non-sustainable carpeting (which can contain formaldehyde and other noxious materials).[94] This creates hazardous off gassing and indoor environmental quality (IEQ) conditions.

---

[93]Name withheld for confidentiality reasons.

While working at the New York Indoor Environmental Quality Center, Inc. (NYIEQ), the team focused on economic and scientific development of technologies to remediate IEQ issues in partnership with the Syracuse Center of Excellence (COE). The company concentrated on funding and commercializing technologies for the built environment that reduced toxins and off gassing inside houses; and on a commercial level in businesses, industry and factories.

However, this should be resolved by repurposing building materials and utilizing low-VOC interior design. Innovations in wind, solar and geothermal that are integrated into net zero housing will continue. Energy efficient housing is the future.

---

[94]Formaldehyde in Your Home: What you need to know. Fact Sheet. Agency for Toxic Substances and Disease Registry (ATSDR). 2018. https://www.cpsc.gov/research-statistics/chemicals/formaldehyde.

Finding New Disruptive Apparel Industries

Three investment presentations that I attended in 2018, one on April 17, 2008, the next on March 21, 2018 and the final on June 21, 2018[95] have included athletic apparel, dress shoes, U.S. based high-end apparel, farm to market and restaurant companies. Direct to consumer startups have been pushed to investors before, including Warby-Parker® and Harry's®; and other subscription as a service apparel firms. There is no indication that funding in this sector will decline, particularly at the seed stage level. However, for apparel startups it is difficult to disrupt established incumbents like Nike® that is nearing $50 billion in revenues. The same is true for clothing, farm to market and restaurants. The economies of scale that established farming operations, restaurants like Nobu Global® and clothing endeavors have are formidable. There is always room for new entrants but distinguishing a company from the competition is significant, especially where price points and profit margins are significantly higher in the new entrant area.

---

[95]The names of the companies are withheld for confidentiality reasons. The venue and the hosting fund names are also not disclosed.

Fleet Logistics and Data Analytics

One of the corporations that I was asked to invest in[96] entered into a data partnership with Oracle® This organization doesn't share information with 3rd parties at all, particularly in light of GDPR and other heightened privacy issues. The data is private and confidential and utilized in aggregate by through the use of API's. The enterprise does compile information the number of aggregated trips that its users make.

Separately the business also has a partnership with Johnson Controls® and is able to predict the life of a battery within 85 to 90% accuracy (60% is the industry standard). The firm's Drive Smart program is based on the idea that safe drivers (i.e. those that have opted into the program) are willing to share their driving habits receive a discount on their insurance.

---

[96]Name withheld due to confidentiality reasons. Based on discussion with startups Co-Founder and CEO on 6/29/18.

The endeavor is the 3rd phase of product development for a fleet management service that allows trucking companies to log their drivers' hours and deliveries more efficiently (via an Electronic Logging Device (ELD). This product is U.S. based only at this time.

It also is involved in the development of a multi-owner vehicle product. This service offering involved 1 to 100 users and costs $25 to $100 per vehicle and is a source of recurring revenue. It also provides an opening to monetize data in an anonymous method. It's partnership with Nielson® is a direct manifestation of this.

The start-up, as a part of its current fundraising round, across Asia, Europe and the U.K. seeks to further expand its operations and sales both domestically and internationally. There is demand for the endeavor's services outside of the Untied States. The corporation has agreements with Audi® in Europe as a part of their co-investments. There is a significant prospect to generate profits in this market, based on existing and projected revenues.

The corporation's service offerings are an opportunity for repeatable recurring revenue, with a low overhead the company is able to operate in a lean fashion. The corporation has been in discussions with UPS® (existing fleet) and Amazon® (developing fleet utilizing Alexa® in vehicle); and possibly integrating with Sidewalks Lab® by Google®. The business views its product and service development flow as tech, automotive, logistics and then packages.

NYC Restaurant

My due diligence on a restaurant includes a company that has secured $200K as a platinum private partner. The restaurant is a chain in New York and is seeking to expand through crowd funding. The restaurant founders have raised capital through investment banking, Wallstreet bankers, private partners and banking loans. Accredited investors (in addition crowd funding) provides an interesting proposition for investors to own a stake in the eatery and act as brand ambassadors (40 to 50 people + associates) for a restaurant that they also may refer friends to, dine and invite their friends to. An investor over the $100K amount receives a plaque noting that they are a significant founder in the venture.

The business has 50K of active users in its email database and a person who acts full time as a social media contact and responds to all email, marketing and social media inquiries.

The investment opportunity is significant: after the 3% management fee, investors will receive 100% of their investment as soon as payback occurs. They then receive 40% of the profits for the time thereafter.

One concern with the neighborhood on the upper West Side of NY, although it is filled with affluent families and does not contain a high-end Mexican restaurant, it does have some vacancies. These vacancies include a Subway® sandwich and other empty buildings. Subway® and its franchisees have been experiencing a slowdown with over 500 locations closing. This may be in part due to the cyclical and fashionable nature of chains and brands (i.e. some brands fall out of favor). However, there is a concern that this restaurant will be impacted if some commercial and retail spaces are struggling in the surrounding neighborhoods.

As an investor it is worth noting the economic climate of the surrounding neighborhood, not just the economics of the venue and the chains history.

From a strategy perspective the restaurant owners conveyed the following:

1. If the restaurant doesn't do well the founder's first desire would be to change it; and
2. The second option would be to sell it – returning any profits to investors.

From the founder's perspective it is analogous to buying a new car. All of the permits are in place and the owner's have a proven track record of success with their other format styles.

At a very specific level Mexican food in particular, is more profitable than other types of restaurant formats. For example in Mexican food, for every $1 spent, costs are only $0.25, compared with other cuisine types that cost $0.35 for every dollar spent. Profit margins are significantly higher as a consequence. Additionally consumers, particularly in New York, are willing to spend $18 to $20 dollars on drinks. The new restaurants beverage programs will be equally well thought out. This will result in a larger profit margin based on the typical check size of $48 /$49 dollars. The restaurant expects to sell 350 to 400 dinners on the weekend and 200 to 300 during the week. It expects to serve over $100K in the winter and significantly over this amount in the summer due to extra seats being available outside.

In terms of the real estate it has a special agreement with the landlord for the space. The landlord is a fan of the founder's chain and has offered to lease the building to them at a lower price per square foot.

The owner has made many basic improvements at his own cost, including improvements to the heating, air conditioning and sprinkler systems. The improvements have totaled $135K in total. The restaurant plans on securing a convertible note of $400K for other portions of the facility build-out. This note would convert to 4%, which is reasonable based on current market rates.

Conclusion

An investment in this conglomerate is recommended based on the restaurant founder's extensive experience and success in the marketplace. The team is exceptional and they have delivered in the difficult NYC marketplace over and over again. The institution has strong ROI metrics and profit margins. This group presented one of the most compelling opportunities of the year.

Detailed Examination: The Rise of Targeted Apparel

This firm (and others that are similar to it) views its branding strategy as more than just apparel. They are trying to build a lifestyle brand like Patagonia®, with styled products that have a much higher price point $55 for unlined shorts and $65 for lined shorts and therefore much higher margins based on the cost of manufacturing in China. The higher margins may also be viewed as a potential long-term weakness, particularly when combined with the fact that the products are manufactured in China, and there has been a trend for "Made in the USA" manufacturing throughout 2017 to 2018.

The company wants customers and stakeholders, including investors, to view their strategy and core mission as providing an athletes' core training kit[97]. They are positioning themselves with established retail players, including Equinox® and Nordstrom®, having resonated with customers. They are actively seeking to position themselves in digital pop-ups, high-end fitness studios and personal trainers, rather than with celebrity endorsements.

---

[97]Discussion with the start-up's founder and CEO on 6/26/18. Name withheld for confidentiality reasons.

The start-up hasn't done a lot with celebrity athletes, like professional football players, instead focusing on brand ambassador programs.

The firm utilizes a template driven landing page (via WordPress®) that can be customized for fitness studios and other corporate clients. Part of the startups' current fund raise includes hiring a Junior Sales Associate to push these template-based packages to high-end fitness centers and other users. This can be considered a strategic weakness given the spread-out geographic location of boutique gyms and personal trainers, requiring extensive travel (which increases costs) in lieu of just phone and web-based CRM acquisition, increasing travel costs to and from these locations. A successful sales strategy includes on-site visits.

The founder's background in technology incudes familiarity with underwriting Everlane® and Warby Parker®, but no skill-sets building a fashion or apparel brand. He built his own brand based on his experience with these previous starts-ups.

The business has a 12% rate of return on its products and has achieved $100K in revenue in June of 2018.[98] The entrepreneurial endeavor is not expecting to raise additional capital, based on the current revenue rates. Fixed costs are low. It utilizes a 3PL SimpleGlobal® that is based in Delaware and whose structure is different from other logistics providers: it scales with volume and provides custom packaging. The firm has warehouses in California and Europe. The costs are $1.30 (1 item) to $1.60 for (2 items) and shipping ranges from $5 to $6 dollars, based on location and rate. Manufacturing is completed in China, relying on this location due to banding and laser cutting machines that are more up-to-date than similar tools in the United States or Europe. The challenge to relocating manufacturing to the United States is that this country doesn't have the tools to support the banding and laser seam capabilities that China does. Margins are approximately 35% (with shirts being higher and shorts being lower). There is an import duty of 30% on the products that the firm's freight forwarder manufactures and distributes. The start-up is also sourcing factories in

---

[98]Electronic communication from CEO & Founder dated 7/6/18.

Columbia (South America) where there are no import duties. There is also a push to relocate the fabric sourcing from a location in Taiwan to one in North Carolina.

The corporation's designer lives in London. She has understanding of a major cycling firm in the UK. They are also seeking to hire someone in the New York, NY to build-out their U.S. design experience. Current fundraising (if successful) will seek to add this person to their roster.

The company is also piloting a trial period where you try the product for 30 days. There is a 4x return on revenue based on overall apparel production volume, manufacturing and operational costs.

The business doesn't have dedicated eBay® or Amazon® sales channels and currently doesn't intend on devoting time or resources to developing a dedicated presence and web pages on these platforms.

Conclusions

The author conducted a test of the apparel (three pairs of shorts, shirts and compression shorts). The rigorous trial involved over 200 miles of running, which revealed the strengths and shortcomings of the apparel. The shorts are the leader in terms of durability. The shirt line is of mixed quality with the cotton shirts being the best constructed; and the synthetic offerings suffering from poor stitching (it has separated and loosened over time). The designer has some work to do compete with the larger incumbents like Nike® and Adidas®. A large positive is several of the shirts have the made in the USA label.

From an investment perspective an outlay in the $25 to $50K range would be conditional: resolve product deficiencies to ensure that the company doesn't suffer from reputational impairment. This should require minor correction on the part of the start-up team.

Footwear

The firm is seeking to become a $100 million revenue corporation.[99] It is expanding its product line from hand-painted dress shoes (at the high end) to sneakers and a resort collection (beach wear and sandals). The start-up wants to be a customer's footwear choice 7 days a week. They use FedEx® for shipping and have a 3PL that is based in Connecticut for logistics. They have had customers from 100 countries, over the lifetime of the brand. However the majority of their customers are domestic in the United States, representing 85% of their sales and revenue. International sales account for 15% of the organization's revenue, with primary locations being English-speaking locations, including Toronto, Canada, London, U.K., Sydney, Australia, Hong Kong and Tokyo, Japan.

---

[99]7/2/18 discussion with the start-ups Founder & CEO. Name withheld due to confidentiality,

The CEO does not anticipate expanding into apparel and intends to focus on profitability and expanding their resort collection. He plans on having 3 to 5 shoes under this product line, which includes, but is not limited to, sandals, sliders, joggers and driving shoes.

The brand is not a freight-forwarder and imports fully finished goods (from Italy). Duties are 8.5%. The freight-forwarder handles all aspects of the manufacturing process from the Italy to the United States and then at the warehouse fulfilling orders. The start-up would receive a benefit from an Italian exit from the European Union (EU), in terms of a change in the Lira's exchange rate in relationship to the U.S. dollar. This would be reflected in the freight-forwarders labor costs reflecting an increased per unit profit margin based on overall revenue rates.

Conclusions

This company is a recommended investment based on the endeavor's solid profit margins. The quality of the product is exceptional. The challenge is distribution. Based on the risky nature of disrupting the incumbents' strong distributions channels. An investment in the $75K range would be warranted.

Other: Televised and Celebrity Hosted Investment Shows

Mark Bennett, the CEO of SharkTank®, takes a 33% equity stake in presenting corporations that are accepted into the show. He also requires that pitching firms pay a $40K entrance fee to be included in the show. This excludes any further equity positions that pitching startups allocate to one of the celebrity investors. Of the approximately 1600 businesses that have presented many that receive a verbal promise of investment, by one of the celebrity investors in the pre-recorded filmed session, do not receive the investment based on follow-up due diligence that is conducted.

A business that is accepting funding, in many cases, is giving away nearly 60% of their endeavor for an investment that ranges in size from $50K to $1 million. Founders should weigh the risks and benefits of losing this much control of their companies at such an early stage of their lifecycle.

Conclusions

They should ask the question: Is the publicity and advertising significant enough to warrant giving away a sizable portion of their enterprise. The celebrity investors and the CEO of Shark Tank® as board members in the firm's that they invest in, hold significant voting power. This can force founders to make decisions that aren't in their own best interests, due to their lack of company jurisdiction. Although the publicity gains for this type of experience are valuable, I would not recommend this investment path for most founders.

Equipment as a Service (EAAS)

Equipment as a service (EAAS) has emerged as a trend for venture capital. It is difficult to compete with Catepillar® and other established players. However, at a June 21, 2018 investor presentation, followed by a June 26, 2018 conference call, there is significant evidence of new entrants in this area, with exceptional promise in the infrastructure, mining, oil and gas and aerospace and military. The founder's family has been involved in mining & gas for several generations. They have signed an exclusive agreement at the equipment producing level with Quantas Services® for power line infrastructure as a part of their EAAS model. The business is targeting Europe and the Middle East for the lucrative geothermal market, as an independent supplier. The enterprise offers demonstrations of their technology at $2,000 per session. Businesses are then contractual obligated for significant shot material consumption moving forward.

The composite material of tungsten and plastic allows for companies to engage in low cost digging, with potential applications for Department of Defense.

The firm is raising a $1.9 million Series A (under Reg A) round due to the $1 to $2 million that was spent on the initial pilot exploratory phase with Shell®. The corporation is valued at $10 million.

The start-up has a proposal with the Department of Energy (DOE)'s DARPA program and has advanced through the proposal process with its co-proposing firm that is based in NJ. The company is also exploring work with NASA due to the scalability of the technology. It can increase the efficiency of a rocket launch by 2km by utilizing the technology in a long tube. Shortening the system's tube makes it weaponizable and/or more applicable to mining and geothermal. However, the firm doesn't have a lot of experience with government procurement and is focusing on the lucrative geothermal market over the long-term.

In Hawai'i, for example, utilities will pay $0.20 to $0.30 per kWh. In Japan there is significant interest in this type of energy source due to its close proximity to the ring of fire.

There is significant risk to this enterprise due to adoption rates of geothermal in the United States, which have not become a standard part of new house construction to date. This inflection point may represent a turning point in the early adoption of this technology for geothermal heating and temperature regulation.

In the transportation, aerospace and renewables segment there is a heightened need for efficiency in the urban transportation sector with access to public and private financing (bonds, et. al.) is critical. Presently this is an overlooked area of venture financing, but is beginning to generate traction with interest in AgTech, Equipment as a Service (EAAS), Aerospace (SpaceX®) and Military Tech and now the broader transportation category led by Hyperloop® (with co-investments by Richard Branson), Tesla® and others.

The subject company is involved in space research through NASA, military and mining operations. It has working agreements with Shell® having demonstrated success in expediting mining operations. It intends to focus on licensing its technology in the near term for rock demolition, foundations and pilings in construction and mining. Catepillar® has an agreement with the firm as of 2017.

The corporation has low operational costs ($10 per firing) and the cost per individual unit is low ($1) as an EAAS. It has the potential to move large-scale infrastructure projects forward at a much more rapid pace (New Jersey/New York, BART/CalTrain, Chicago - Hyperloop®).

Conclusions

Risks for the enterprise include high procurement barriers to entry (as examples: MTA, Chicago Transit, BART; and other international regulations). Benefits for developing large-scale projects (infrastructure, tunnels) are exceptional: reducing the time and speed of mining and underwriting costs (this is one of the extraordinary expenses attributed to large scale infrastructure). An investment in this enterprise is a go. Despite the procurement barriers to entry, the profit margins of this company are large. This is a firm to watch out for in the future.

Biotechnology and Biomedical VC in 2007-2009

Biotechnology and biomedical companies have utilized venture capital to fund their operations at a variety of stages. In 2007, venture capital investments in California's life sciences (medical equipment and biotechnology) "increased to $4.3 billion from $3.2 billion".[100] Additionally:

---

[100]California Biomedical Industry. 2009 Report. California Healthcare Institute (CHI) and Pricewaterhouse Coopers, LLP. 2009.

The number of deals increased to 315 in 2007 from 299 in 2006. Biotechnology deals alone garnered $2.3 billion in 2007. That figure marks an increase of approximately 21.7 percent over the $1.9 billion raised in 2006.[101]

Biotechnology venture capital investments have increased substantially over the last couple of years, particularly in California which as of 2007 had the largest "venture capital investment in life sciences by state" at approximately $4.4 billion.[102] In general, in 2007, California entrepreneurial endeavors "continued to draw more capital infusions than did life sciences companies in any other state" at $14.7 billion out of a total of $30.7 billion.[103] California can be used as a benchmark for gauging venture capital investment trends.

---

[101]*Id.*
[102]*Id.*
[103]*Id.*

Biotechnology and Genomics VC Now

Genomics (including DNA analysis) is one biotechnology and science focus area that has received funding. As noted previously, in 2017, Genomics companies received record funding in 2017 at $2.5B.[104] DNA sequencing companies can expect to continue to receive funding in 2018 and beyond. However, at a certain point this market will become saturated and investors will move onto another sector.

Conclusions

One of the limitations of the medtech sector is that many of the major conditions and diseases have yet to be resolved: these include, but are not limited to, multiple sclerosis (MS)[105], Parkinson's disease, HIV, et. al. Many of larger pharmaceutical companies could devote more time and energy to tackling these larger societal problems. A startup fund devoted to any one of these conditions and run by a large pharmaceutical corporation would add great value to the ecosystem. There is a tremendous value proposition for investors to tackle these systemic issues.

---

[104]PwC/CB Insights. *MoneyTree Report™ Q4*. 2017.

Medical Device Companies

One of the medtech businesses under my review was a startup developing a usb enabled device for women (and lesser extent men) facing issues with their urinary tract (and in some cases requiring surgery for incontinence) due to leaking urine after child bearing.

Female and male incontinence is the #1 issue and/or reason for entering a nursing home. Why should this be the case when such great technology exists? It shouldn't.

The device sends currents from one portion of the body to another – wearable while getting treatment. The device has a removable gel pad that is disposable. Gel pads are used 4 times a week; and include 5 pads in a pack. The device is custom made on the founder's 3-D printer. The cost of goods sold includes the electrodes and packaging and represents approximately 25 to 40% percent of the overall cost ($500 x .25 = $125.). Hologic is involved as a 3rd party in the companies manufacturing and distribution process.

---

[105]My funraising activity includes the following: MS Cycling and running events in California and Connecticut.

The Food & Drug Administration (FDA) submitted an October 8, 2018 De Novo letter with specific clarifications and demands.

The brand has responded to this inquiry in January, but still doesn't have a formal reply. The device is ISO 13485 Certified.

The device has been undergoing a self-administered clinical review with sixty (60) subjects for two-weeks (including a placebo) and a usability study in Fort Wayne, Indiana for twelve weeks (12) from August to September with five-hundred (500) women. Additional public fronting events have included tradeshows. The company is seeking FDA approval in order to obtain Medicare co-payments, having established valuing metrics based on Facebook® advertisements (split testing at $250, $480, $600 and $780: $600 was clicked on the most) and one-hundred (100) interviews.[106]

---

[106] 10/15/18 discussion with Co-Founder.

Pricing for each unit initially was $725 dollars COGS <$60 – profit margin - 80 to 90 percent (this price has subsequently been reduced to $500 based on studying women's Willingness to Pay (WTP), doctors input and tradeshow feedback; and once the business lowers its prices Medicaid reimbursement amounts will follow.

As a consequence the company is seeking to price its unit with as high a margin as initially possible to avoid Medicare from reducing its own reimbursement level. The unit is available over the counter and via prescription if the user is seeking reimbursement.

It is important to note that the Department of Defense (DOD) has developed a male electrode device. This enables soldiers and other family members to benefit from this technology in the field, in the air, upper atmosphere, outer space and elsewhere.

As with many earlier-stage medical companies the risk lies in FDA approval. I proposed several follow-up questions, as follows:

1. Exit-strategy - Is the firm focused on an M&A acquisition; or longer-term growth with expansion of its product pipeline - i.e. male device

(recognizing that the team indicated this is not a focus area initially); and/or other product silos?

The firm acknowledged that in the medical device area, acquisition is the typical scenario.

However, they noted that they are planning for long-term growth. Specifically, they noted:

> "We do think growing the product pipeline is relatively easy since it is based off the same technology. We also believe that if we have some sales and another product in the pipeline the valuation will be substantially higher. We are actively talking to strategic partners such as Medtronic and Hologic. In our eyes, we are hoping for a sales partner since they have a strong salesforce to gynecologists."

The company doesn't have anything that it actually developing in the product pipeline. This is a significant problem for reaching the market saturation threshold. I wouldn't invest in this enterprise until they launched their concurrent male device (part of a NIA grant); or licensed the application for other developments.

Willingness to pay (WTP) - To clarify the pricing strategy (via the social media and other outlets where the range of potential cost per unit was introduced) - As the devices are not sold directly at CVS®, Wallgreens®, Duane Reed®, Rite Aid® and/or online - Is the WTP expected to be the same in stores versus online (i.e. differences in distribution, inventory, et. al.)?

The firm has projected that the WTP is higher for the convenience of ordering a medical device online. The start-up specifically noted,

"Willingness to pay is slightly higher online for discreteness and convenience. We have interviewed women about this and they do not want to go to the store. This is an embarrassing problem. I think inventory and distribution then becomes easier for us and we don't have to pay for shelf space and management of that. We could easily utilize our partner Arrow Electronics for manufacturing, inventory management, storage and shipping."

The concern here is that company pricing strategy will become fragmented through its online and in store channels. Consumers are often embarrassed to buy birth control, morning after, condoms and/or other products in the store.

However, physical locations are still an important venue for achieving significant sales. In my direct observation, in store traffic in drug stores is still very high.

2. If FDA approval is not granted (although team noted current clarifications are minor) - How far potentially does this shift the formal product launch?

The brand's claim that 1 in 3 women experience this issue is a valid claim. However, the actual likelihood that women will pay $500 for the device without Medicare and/or substantive insurance co-pay is unlikely. This creates a very niche marketplace of customers with an actual willingness to pay (WTP).

From an acquisition perspective this makes a merger & acquisition (M&A) strategy the most likely scenario. The valuation is most likely lower in this case because the company's product pipeline is not very well established. Risk threshold is very high due to current lack of FDA approval. From an investment perspective it is a pass at this juncture.

Conclusions

Investors should continue to expect more bluetooth and usb enabled medical devices that are connected to smart phones and can upload data onto a computer. This trend will only grow more exponentially, following the surge of sports wearables. It is an attractive investment area, providing high profit margins.

VC's can be expected to continue to invest in insurtech, compliance and other regulated industries. There is a huge need for these types of firms and the fragmented services that they currently supply. The upside of cost savings and efficiency in this sector are noteworthy. Expect companies and services in this sector to continue to receive substantial funding rounds.

Issues VC Industry Faced in 2007-2009

A National Venture Capital Association and PricewaterhouseCoopers press release citing the PricewaterhouseCoopers *MoneyTree Report* for the first quarter of 2009 notes, "Venture capitalists invested just $3.0 billion in 549 deals in the first quarter of 2009".[107] Additionally, this press release indicates "Quarterly investment activity was down 47 percent in dollars and 37 percent in deals from the fourth quarter of 2008 when $5.7 billion was invested in 866 deals".[108] This overall total was reflected in the corresponding decline in individual sectors and subsectors of the venture capital industry.

---

[107] *Venture Capital Investment Plummets in Q1 2009 to 12 Year Low.* Press Release Citing PricewaterhouseCoopers. MoneyTree Report. Data provided by Thompson Financial. PricewaterhouseCoopers, LLP. April 18, 2009.
[108] *Id.*

Conclusions

The industry experienced a downturn based on the recession at the time. However, the investment community still sought to deploy capital during this period. This is reflected in the $3.0 billion dollar deployment despite the overall downturn of the market. It demonstrates that investors were still attracted to risk during a period of high economic uncertainty.

Major Issues Facing the VC Industry in the Near Future

Venture capital partnerships are likely to be more focused and exercise greater due diligence in their investments in AI, VR, Fintech, Cybersecurity, Blockchain, biotechnology, cleantech, software and Web 2.0 (now generally defined as SAAS and PAAS) technologies in 2018 and 2019.

One of the larger issues facing the industry is what new sectors will emerge in the U.S. economy that will attract the attention of venture capitalists. It is likely that AR, VR, Blockchain, biotechnology, cleantech and software (SAAS and PAAS) will continue to receive large allocations of private equity, even with funding declines in most sectors.

New Mexico and Arizona are regions that have experienced rapid venture capital growth in "sectors of growing interest to venture capital investors - such as energy, advanced materials, water treatment, optics, and high-performance computing".[109]

---

[109]Clare Chachere, PricewaterhouseCoopers, Emily Medell, National

Safety is one of the largest issues VCs and the industry is facing. Technologies like Uber®[110] and Tesla® have had fatal crashes with cyclists and the highway median.

The industry is moving too quickly on self-driving cars and unproven biotechnology DNA sequencing and testing methods with Theranos®, Wallgreens® and the FDA violations case.

Cybersecurity issues threaten the nation's banks and utilities; and operations of government and are completely out-of-control and require drastic remediation measures, including direct government intervention.

Venture Capital Association and Porter Novelli for PricewaterhouseCoopers. Fastest Growing Regions for Venture Capital Investment Lie Outside Silicon Valley. Venture Capitalists Finding Opportunities in "Unexpected" Pockets. National Venture Capital Association (NVCA). PricewaterhouseCoopers. New York, March 11, 2008.
[110]Troy Griggs and Daisuke Wayabayashi. *How a Self-Driving Uber Killed a Pedestrian in Arizona.* The New York Times. March 21, 2018. https://www.nytimes.com/interactive/2018/03/20/us/self-driving-uber-pedestrian-killed.html.

Accounting Strategies for Hedge Funds, VC's and Start-ups

It is imperative for hedge funds, VC's and start-ups to deploy several accounting firms based on the stage of their firms and/or businesses. One accounting firm that I met with does just this: They provide accounting and compliance services for hedge funds, start-ups and other companies, including 1020 and 1120 filings (which are standard and are required).

The enterprise in many cases recommends having start-ups convert their structures from and LLC to a C-Corp; and/or a C-Corp to an S-Corp. With a C-Corp in particular if the founder's hold their investment for 5 years they can obtain a full exclusion, in some cases, from the IRS. These benefits for firms with revenues of less than $50 million can include exclusions of greater than $10 million or 10x. An examination of the 1202 deferment of capital gains in these circumstances is critical, particularly for start-ups providing technology services, SAAS as a qualified C-Corp. There is often research & development (R&D) credits that qualified high-technology businesses may be able to access. 605 early adopter benefits are also possible. The firm can determine if deferred revenue is too big for conversion and/or what the deferred revenue schedule is. The partnership has a standard non-disclosure agreement (NDA) and can provide expertise on the capitalization of software schedules, a review of cap tables, time & material (T&M), operating expenses and ending capital reserves.

The accounting firm can conduct a review engagement for process improvements for minimal cost, or approximately $15 to $20K. They can complete these bookkeeping services in the off-season (after April) for significantly less. A larger audit costs $50 to $100K and includes validating contracts with their attorney.

Fees for this organization to act on a retainer basis are fixed and are on a base fee level. Additional costs for Opinion Letters, for special determinations and rulings from the IRS are extra.

Conclusions

Having a second firm on retainer is valuable for any startup or established business. They can tackle growth issues. Smaller financial management companies don't have the experience supervising larger accounts, or the knowledge of the complexities of Irish and/or overseas tax law. Start hunting for these services before you need them. Then you can make the call and they will be in place before your start-up has a crisis.

Seed-Stage Investment Network

My entry into a new early and mid-stage investment network and platform occurred in 2018, specifying interests in fintech, biotechnology and cybersecurity (as areas of investment focus). This excellent investor network[111] screens about 50 companies per year. Several of the companies "leave" the network due to positive disruptive reasons, including mergers & acquisitions (M&A) transactions; and joining incubators (or accelerators). The network is free to join and sources quality deal-flow. In its current portfolio are 6 to 7 healthcare companies, with some focus on biotechnology, healthtech, medtech and other SAAS based multinationals. The VC entity at its current stage includes two very early stage cybersecurity organizations.

On the institutional investment side the conglomerate hosts several hedge fund, family office, fintech and blockchain focused events. These conferences are held in New York, Boston and on the West Coast in California.

---

[111]Name of investment network withheld for confidentiality reasons.

Conclusions

Many of these smaller entities provide access to interesting deal flow that larger institutional VCs are unable to touch. Several of the deals that I have reviewed are focused on labor compliance, a comedy sourcing platform, a restaurant expansion, medical devices and even apparel. This provides both great occasion and risk for accessing deal flow at an early stage. However, one of the limitations of these early stage networks is that they have a limited amount of participating companies and many lack fully developed cybersecurity, biotechnology and fintech start-ups, due to pending regulatory approval.

Mergers & Acquisitions (M&A)

I was approached regarding a Merger & Acquisition (M&A) deal.[112] For the groups conducting transaction due diligence, initial fees range from $8,000 to $10,000 for the first six months; and $1,000 to $3,000 thereafter. In this specific case the two-tiered fees and structure were $8,750 for the first 6 months and $1,250 per month thereafter. This equates to $8,750 x 6 or $52,000 + $1,250 x 6 or $7,500 thereafter for a year total of $60,000. This serves as an example of the math. The actual time for deal closure in an M&A transaction and deal may be shorter or longer. The base fees need to be adjusted accordingly.

---

[112]Confidential discussion. Names withheld.

4 to 10 people are assigned within the M&A advisory team. Due diligence from this team is separate from the internal and external attorneys that may be needed to close the deal. The M&A team selects 3 A-level buyers where there is an obvious fit, 10 B-level acquirers; and up to 750 potential companies overall, including outliers. The M&A team researches the strategy of the buyer (and/or buyers) and consults with valuation experts (essentially appraisers). If there is a successful transaction then an additional success fee of 4% to 5% of the total transaction is applied. For example if the deal is $10 million then the success fees are $400,000. This may come in the form of a cash or stock payout; or other means of compensation.

There are typically no additional filings (i.e. like a Form D, which is used for fundraising and filed with the SEC). However there are selling documents and a letter of engagement that is prepared. There approximately $110,000 to $130,000 in fees associated with this stage and part of the M&A deal structure. The letter of engagement is coupled with a Non-Disclosure Agreement (NDA) and the release of information is staged.

It is at this stage of the transaction the deal size might be negotiated down. For example, during discovery the buyers may raise issues with revenue rate or profitability; or growth and offer to pay less (i.e. $9 million instead of $10 million). M&A transactions have focused on the standard SAAS, PAAS, biotechnology, fintech and insurtech companies in addition to Initial Coin Offerings (ICOs), bitcoin and blockchain. M&A firms also complete Reg D filings; or may in some unusual situations participate as a part of larger team in an initial public offering (IPO). However, the subject firm's primary focus is M&A and fundraising.

In either an M&A and/or IPO it is possible to retain some assets if the deal is structured correctly. In this case a shell company would retain the assets that it wanted such as trademarks, patents and technology through the shell corporation and sell the portions that the buyers sought.

Regardless, for a PAAS or SAAS the end-use license agreement must be revised and the parties notified in the event of a change of ownership. This requires more than just a letter or phone call but in most cases direct contact with large enterprise users.

After the anonymous high-level of intent documents are prepared, a 6 to 10 page executive summary outlining standard metrics is prepared (competition in the marketplace, shareholder lists, disclosure documents, number of staff and other standard due diligence items). Then the M&A firm sets up a direct call with the buyer, ensuring most of the protective elements are in place for the seller and buyer's due diligence and review (including Weighted Average Cost of Capital (WACC), companies in a similar 1/3rd range of the seller's peer group and their respective cash flows; and what is the 12 month outlook for revenue and profitability and beyond). This type of information is released very late in the discovery process.

Additional steps in the due diligence processes include a review from the seller's perspective: what is the process for integration (which ties into organizational culture) within the buyer's company; speaking to another business that has been acquired; evaluating the buyer financially; and other pertinent factors.

The CEO and legal team complete a full seller's due diligence worksheet, including every aspect of the firm (retention rates, vertical sectors, competitors, articles of incorporation, employee NDA's and any other aspects of capturing the range of the businesses operations. Many of these elements are also included in the IPO process. Prior funding rounds are typically not taken into consideration, i.e. A-D rounds, and/or even bridge funding.

At this stage of the M&A process potential roadblocks to the completion of the deal, both major and minor begin to appear. Once the term sheet has been signed, the level of leverage goes down significantly. In fact many deals may be cut by 25% or 50% from $10 million to $5 million based on prior example.

Conclusions

M&A is not for every corporation. In some cases an IPO may be a more advantageous path: M&A is one route, but it has an end, in the sense that the startups founder and other employees may complete their legal obligations (non-complete clauses) to the seller and leave to start a new venture after 1 to 5 years depending on the structure of the deal.

Canadian M&A and Advisory Groups

There are companies that provide similar M&A and advisory services throughout the world. One Canadian listed firm[113] charges up to $13,000 initially, with a 4% success fee (completed in Canadian dollars) at the exact moment the transaction closes.

This firm relies on a number of investment underwriting options, including crowd funding from a pool of accredited investors, venture capital firms and pension funds. Convertible bonds are also an option, in addition to equity raises.

---

[113]Confidential discussion. Names withheld.

Equity raises with this firm typically range from $1 to $5 million and have reach up to $20 million in dollar size. Nearly 60% of the M&A and transactions are based in North America and are focused on the tech space and life sciences. Listing services include filming a corporate video; PR, social media and investor marketing campaigns to their investor base, under the purview of an account manager.

Conclusions

In many cases the risks of an international listing outweigh the benefits. If you are a U.S. based firm you should weight the potential risk and additional legal/tax liabilities, consulting your legal and accounting teams before proceeding.

Organizational Culture in VC and Technology Companies

The issues in the VC and technology industry surrounding diversity and inclusion require a structural behavior study of both specific firms and the industry at large. Technology companies, by design and through the nature of coding, separate themselves from their users.

It doesn't have to be this way.

A 360 evaluation of companies' employees combined with an organizational behavior study and network mapping can identify some of the deep seeded biases that exist in the industry and with individual employees. Why do the technology and VC firms struggle with issues of including women, minorities and other protected classes in their companies? It is because they haven't fully examined themselves.

Companies like Apple®, Google® and other VC backed firms must take proactive steps to address their low levels of inclusion regarding women, African Americans and other groups. This requires more than just releasing diversity reports on a quarterly or annual basis: 360 evaluations, organization behavior and network mapping (identifying influencers) can begin to reverse the deep seeded biases that exist in technology and VC firms.

It is more than just an academic exercise: African Americans, women and other groups are great employees. They add a different perspective. This has tremendous value.

They are also stakeholders in their respective communities and their friends and neighbors are consumers of the very products that the technology companies are making, including, but not limited, to iPhones®, iTunes®, Microsoft Windows® and social media. Walking through New York City and riding the subway, advertising for Apple® products have included African American musicians and women as cornerstones of their campaigns for years.

If this music and culture is good enough to appropriate for selling billions of dollars in products and services, why aren't these groups hired and employed in these same companies in greater numbers?

An examination of the mapping and 360 analyses of employees can begin to identify the roots of this problem. It is absolutely essential that technology companies take proactive steps and invest in these studies of their companies.

The payback will be extraordinary.

Organizational Behavior Mapping of a Joint Venture

Introduction

One of the principle questions that emerged in conducting a social network analysis and organizational context diagnostic is whether or not the network patterns that are being influenced by a firms' formal structure, work management practices, leadership, culture and human resources practices are specific objectives.

For the VC and technology industry this type of study and design can be used to identify corporate biases that are leading to a lack of inclusion and dysfunction.

In an examination of an administrative context diagnostic it was clear that the JV (in Hawai'i that was the subject of this study) was in need of significant improvement in order to become a more functional and productive firm and legal entity.

Organizational Context Analysis of a Joint Venture

In order to procure and be awarded a large military Indefinite Demand Indefinite Quantity (IDIQ) contract a governing mechanism has to be formed[114]. In terms of the actual resource allocations to form this Joint Venture (JV) it required modifications to the lead company's insurance policies to include the newly named entity. It also required the relocation of one California employee and one Virginia manager. In this case the target percentages for each business had been set at 40% for the lead partner, and 30% for the other venture partners.

Organizationally, the three companies had to form a separate legal entity to manage the complex relationships between each of the three primes; and the twenty-five (25) plus subcontractors that could be potentially assigned to each Task Order.

---

[114]Actual Case Study. Names Withheld.

The official legal name of this JV (name withheld) had two managers: the Contract Administrator and the Project Manager who are responsible for meeting the requirements of the President, the Board, the individual companies that make up the JV and the twenty-five (25) plus subcontractors (including small business reporting requirements; and minority and women owned businesses).

This management team creates a unique and difficult organizational design within the parent corporation, which was responsible to a much larger group of companies. Its functions were largely separated from the rest of the corporate office (a specific driver). For example, managers were required to negotiate Task Orders on behalf of the other companies. The negotiation process did not commence until a Project Manager who originates from the parent office had the necessary qualifications to negotiate and bid on the specific Task Order (the Guam Archaeological Survey for example).

For example, if the qualifications for the Archaeological Survey work could not be met by personnel from the main office then the lead partner would need to bring someone from their Boise office to complete the work or transfer the "right" to bid on the project back to the one of the other co-partner companies.

This limited the sharing of information about projects and the overall collaboration within the larger group for the following reasons: because a geographic office was seeking to retain as much business as possible and continue to keep their billable hours high in relationship to other offices.

Collaborative Issues

Regardless of whether or not a office on the mainland (California, Midwest, East Coast or Southeast) secured the work, the executive culture of the Honolulu office in many instances is starkly different from the operations of the JV which creates a disconnect between the two operations (a driver). This disconnect occurs even with at least two managers being located in the lead partner's Honolulu office because the majority of the team's legal staff, Human Resources, Chief Financial Officer (CFO), and President of the JV and the lead partnership were all located in the Charlottesville Office (where the JV was headquartered).

Additionally, all other JV officers were located in other parts of the U.S. (another complicating driver). This structural disconnect occurs primarily because the managers are employed by the parent organization and reside in the firm's Honolulu office. The managers were expected to participate in office functions but are also on-call to answer to the Virginia headquarters, NAVFAC Pacific (the client) and other members of the partnership and their respective subcontractors.

This created a certain degree of tension regarding the existing Honolulu office members and the JV. Specifically, this conflict emerged because unless Honolulu office members were assigned to the JV work (as project managers or planners) the rest of the office personnel aren't involved in negotiation, fee proposal development and contract development (i.e. specifically the business functions of the JV's operations that are being developed by the JV staff in the Honolulu office).

In essence these non-JV employees were left completely in the dark about the operations of the JV, which created uncertainty and some distrust. This created an "ownership" issue for the projects that the JV negotiated and developed for members of the Honolulu office. It also created a "stakeholder" issue for those projects that these same office personnel weren't involved in because (even if they are informed of the JV's operations as an overview, which they frequently are) they still didn't feel connected with the day-to-day operations.

This produced an "us" versus "them" environment in many situations and directly circumvents any potential for collaborative efforts.

Further Collaborative Complications

The separate corporate responsibilities that the lead corporation and the JV have were further complicated by the requirement that all Honolulu (and companywide personnel) were required to have billable hours or be assigned to specific contract work that may or may not be a result of the JV's operations (a specific complex driver). All JV staff members were given a budget of these billable hours that were developed specifically for the procurement work that was necessary to secure a Task Order under the larger IDIQ contract vehicle.

No other staff members were able to use or compete for these JV billable hours. However, existing non-JV staff members had to "compete" for billable hours and attempted to fill up these billable hours to 100%, even as JV staff in most months are nearly sixty (60%) to eighty (80%) percent billable (i.e. meaning that the perception of the staff that are competing for projects to make them 100% billable, while the JV Managers had a built-in billable mechanism).

This shaped significant collective disconnects that occurred between the headquarters and regional office managers; and the JV employees because they were in direct competition without a distribution of billable hours.

It is also worth noting that the partnering companies have their own corporate structure with elements that are similar to "billable" hours (also a driver), which although not the focus of this analysis, further complicated the relationship within the JV and collaborative efforts between the companies involved.

Specific Analysis of the Organizational Context Diagnostic

A review of the diagrams developed for this directorial context diagnostic reveals significant disconnects (see detailed charts below) between the actual effectiveness of a combined formal structure, leadership and culture, work management practices and human resources practices. Both entities had separate formal guidebooks that dictate how policies and procedures are developed throughout the corporation.

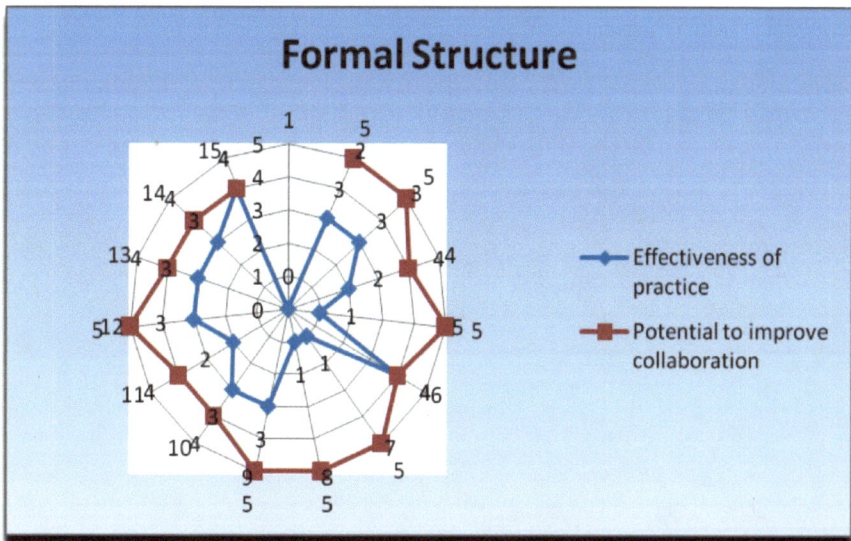

The overall formal structure of the legal entity is flawed with significant gaps between effectiveness of practice and potential to improve collaboration.

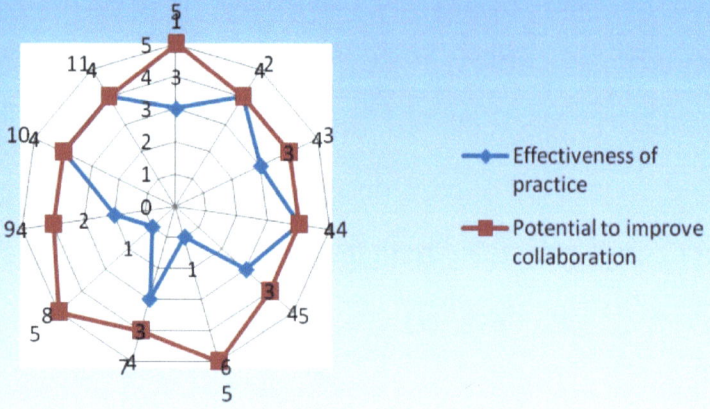

Work Management Practices

Work management practices are generally acceptable with fewer gaps between the effectiveness of practices and the potential to improve collaboration. However, there is room to improve overall processes in this area.

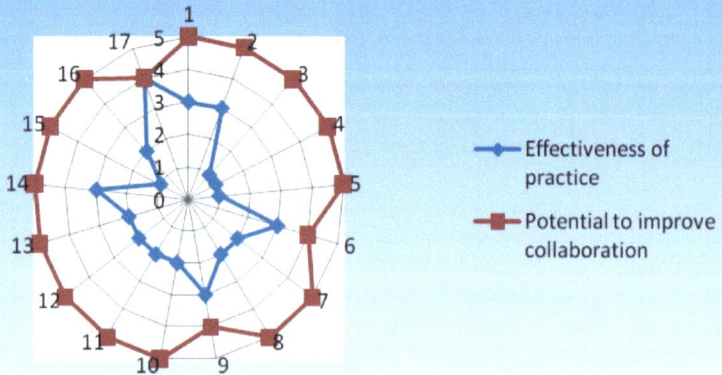

Leadership and Culture

A vast disconnect exists in leadership and culture and perceptions of effectiveness of practice with the potential to improve collaboration.

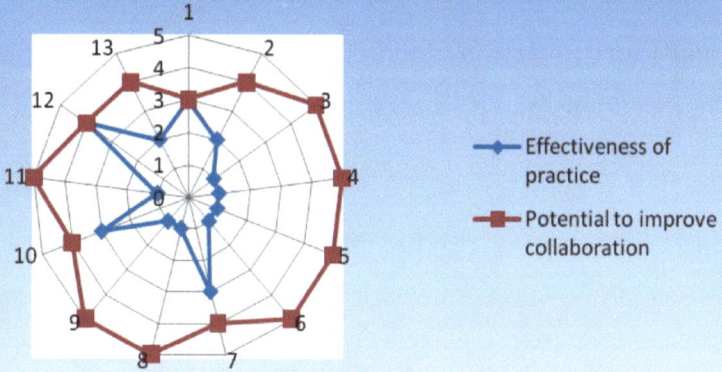

Human resources are in serious need of improvement in most areas. The potential for better collaboration is significant and more efficient organizational alignment needs to occur with the effectiveness of practice.

The actual specific questions and the same charts are included after the conclusion section of the paper.

Conclusion

An organizational context diagnostic provides valuable insights into the actual social and work operations in the JV. The logistic context diagnostic (outlined below) indicates that there is a vast amount of improvement that could occur within the partners.

The study would have been more useful if multiple people within both entities participated in the study to arrive at a more balanced perspective. It would have also benefited from the involvement of employees within the headquarters that directly worked in this unique legal entity as employees. The administrative context problem-solving is only truly useful if the results were analyzed in order to arrive at specific conclusions and form definite new business practices to improve the overall operations of the corporate structure.

In many cases instead of making the appropriate changes management continues with a system that is dysfunctional, even going to the extent of attempting to transfer the responsibility onto individual employees with training courses with titles like *Communicating with Purpose*. These are not a solution by themselves. They do not solve the root of the problem. If a firm's culture has extreme problems that verge on violence (use of slurs, bullying, obscene gestures, harassment and intimidation), then these situations require aggressive internal and external intervention. This requires replacement of management by a company's board of directors; and even the replacement of the board, which has occurred with Wells Fargo®. Management needs to have an audit of the business and include a 3rd party consultant to complete this process in a timely and expedited fashion.

Assessment # 3: Organizational Context Diagnostic for the JV

Effectiveness of practice in promoting collaboration	Potential of practice to improve collaboration
Very Ineffective (1)	Strongly Disagree (1)
Ineffective (2)	Disagree (2)
Neutral (3)	Neutral (3)
Effective (4)	Agree (4)
Very Effective (5)	Strongly Agree (5)
This Practice Does Not Exist (6)	Not Applicable (6)

Formal Structure

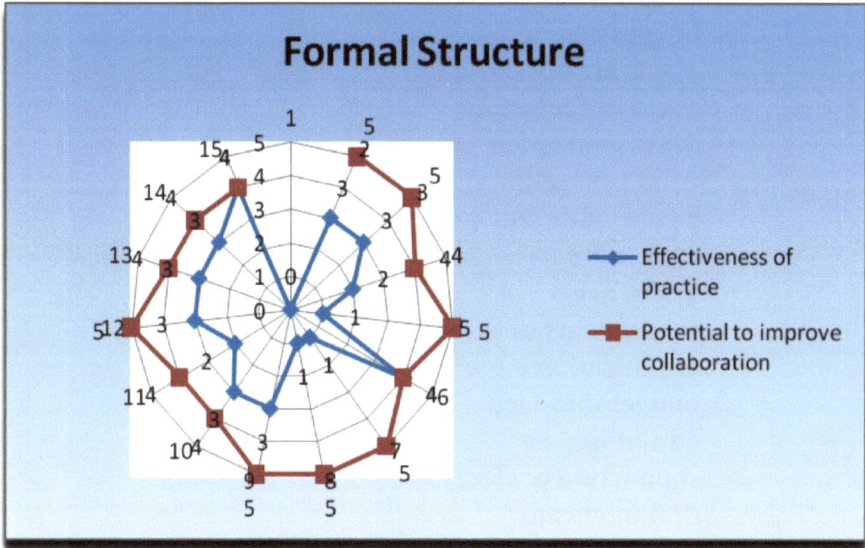

Formal Structure

Formal Structure		Effectiveness of Practice	Potential to Improve Collaboration
1.	People in this network are encouraged to reach out to another function for expertise without going through a formal procedure or chain of command.	3.	5.
2.	Planning processes and goals explicitly address integration of functions or divisions.	3.	5.
3.	Planning processes	2.	4.

	help develop insight as to how integration of disparate expertise could differentiate the organization from competitors or provide value to customers.		
4.	There are components of the organization's budget that focus on funding or supporting projects that integrate people with different expertise of from different functions and divisions.	1.	5.
5.	There are processes and procedures (or accepted cultural norms) that make it easy for one person to reach out to another hierarchical level without going through the chain of command.	4.	4.
6.	People in this network know which decisions they are allowed to make and which they need to consult	1.	5.

	others on (and who those other people are).			
7.	Decision rights are effectively allocated throughout the group so that work is not excessively slowed in order to obtain approvals.	1.		5.
8.	Information is effectively distributed in the group rather than people having to turn to someone at a higher level for information to get work done.	3.		5.
9.	Positions of influence (or committees) in this hierarchy are spread across functions or business units to help ensure integration within and across functional boundaries.	3.		4.
10.	There are specific roles (such as knowledge managers) or pieces of roles (such as modified staffing coordinators) that	2.		4.

	help people connect across physical and functional boundaries.		
11.	There are informal or liaison roles that establish a point of contact for communication between functions or business units within or outside of the group.	3.	5.
12.	Rotational assignments help integrate this group by creating relationships across boundaries created by function or physical space.	3.	4.
13.	Communities of practice are supported in a way that helps integrate networks across physical, functional, or hierarchical boundaries.	3.	4.
14.	Internal initiatives, such as committee work, philanthropic efforts, recruiting and sports, help integrate people in the network.	4.	4.

Work Management Practices

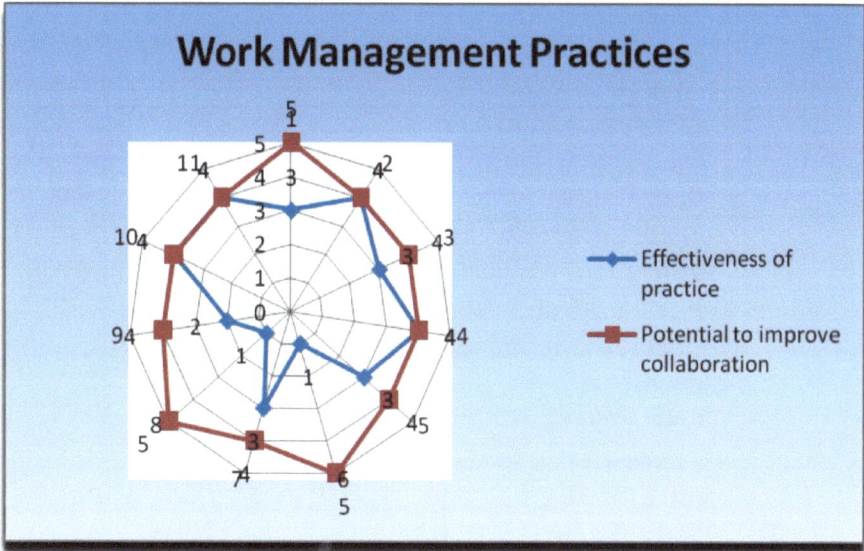

Work Management Practices		Effectiveness of Practice	Potential to Improve Collaboration
1.	The employees with the most relevant expertise (rather than just those whom a leader knows and likes) are assigned to projects when they are initiated.	3.	5.
2.	Once projects are staffed, all employees are encouraged to seek out those with the most relevant expertise (either in the	4.	4.

	group or elsewhere in the organization).		
3.	Employees have enough time to seek input from others or to make themselves available to help others.	3.	4.
4.	People are able to shift tasks to the people with the most expertise.	4.	4.
5.	There are integrated handoffs for products and services that move through different functional areas.	3.	4.
6.	The physical space in which this group is housed facilitates spontaneous communication.	1.	5.
7.	A balance of synchronous and asynchronous technologies is used to support virtual work.	3.	4.
8.	Skill-profiling systems exist that allow individuals to tap into expertise not	1.	5.

	already known to them.		
9.	Synchronous technologies are employed to supplement face-to-face interactions.	2.	4.
10.	Asynchronous technologies are employed that allow people to query others or store work products.	4.	4.
11.	Instant messaging allows for serendipitous interaction.	4.	4.

Human Resources Practices

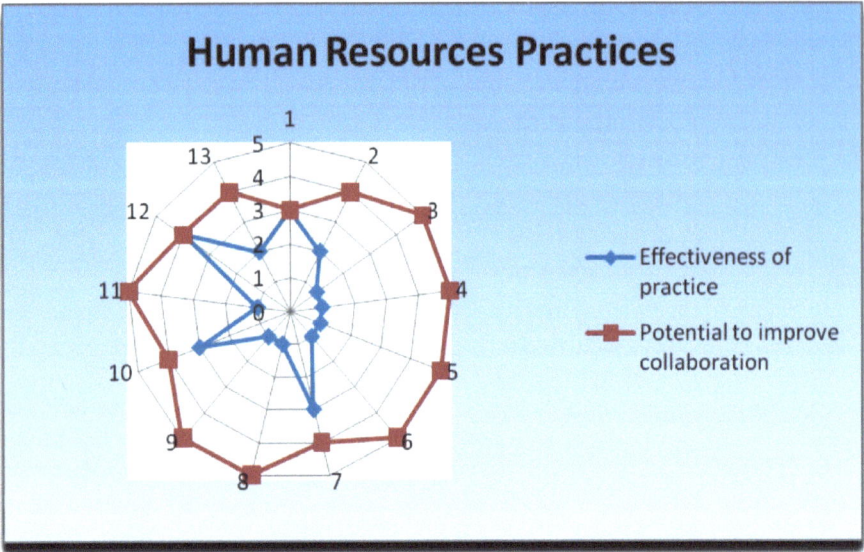

Human Resources Practices	Effectiveness of Practice	Potential to Improve Collaboration
1. This group's recruiting process screens for people who have demonstrated collaborative behaviors.	3.	3.
2. This group's recruiting process screens for people with depth and breadth of expertise, which will make them effective integrators across disciplines.	2.	4.

3.	Orientation practices help new people develop an awareness of who does what in the organization.	1.	5.
4.	Orientation practices help make the group aware of the new person's expertise.	1.	5.
5.	Efforts are made to conduct orientation in groups so that new people have a network right from the start.	1.	5.
6.	There are activities to support new cohorts after orientation, such as ongoing training and informal get-togethers.	1.	5.
7.	In general, there is an effort to conduct training in a group setting rather than sending individuals to customized programs.	3.	4.
8.	Professional development plans help individuals develop their personal networks.	1.	5.
9.	Demonstration of	1.	5.

	collaborative behaviors is a meaningful component of performance evaluation.		
10.	Performance feedback (at least in relation to collaborative behaviors) is given by sources that have witnessed the behavior.	3.	4.
11.	In general, the people who get the largest raises or bonuses are rewarded on their collaborative behavior.	1.	5.
12.	This group employs "spot" reward mechanisms for collaborative behavior.	4.	4.
13.	People in this group intrinsically value collaboration as a part of their network.	2.	4.

Leadership and Culture

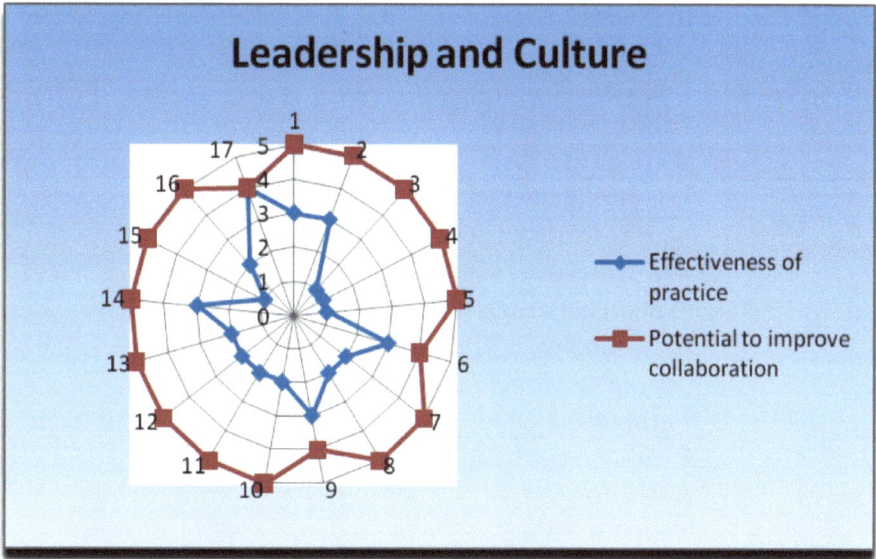

Leadership and Culture		Effectiveness of practice	Potential to improve collaboration
1.	Leaders of this group envision and structure work as a collaborative endeavor.	3.	5.
2.	Leaders encourage collaboration in problem solving.	3.	5.
3.	Leaders focus on involving people who might be on the periphery of networks.	1.	5.
4.	Leaders help	1.	5.

	employees build their own personal networks.		
5.	Leaders are willing to share their networks.	1.	5.
6.	Leaders direct people to those with relevant expertise rather than forcing people to come to them.	3.	4.
7.	Leaders are quick to spot points within a network experiencing tensions.	2.	5.
8.	Leaders or this group are active and effective communicators.	2.	5.
9.	Face-to-face forums are done in such a way that people develop social ties and learn about the expertise of others.	3.	4.
10.	Face-to-face forums are done in such a way that people develop social	2.	5.

	ties and learn about the expertise of others.		
11.	Face-to-face forums are inclusive rather than the domain of a select few.	2.	5.
12.	People are committed to a broad goal and set of values that help promote integration throughout the entire network.	2.	5.
13.	"Stretch" goals encourage people to seek out allies, resources, and solutions across boundaries.	2.	5.
14.	"Unwritten rules" do not prevent people from working across boundaries, sharing bad news with bosses, or admitting failure.	3.	5.
15.	In general, this is a safe environment	1.	5.

	where people are not afraid to admit a lack of knowledge.		
16.	There are sufficient opportunities for people to develop trust in others.	2.	5.
17.	People are willing to share information in a draft format rather than perfecting their work first.	4.	4.

Case Study on Banking Organizational Culture

Organization behavior analysis is not only critical to analyzing multi-firm joint ventures it is also imperative at banks and financial institutions. Included below is a summary of the findings of the financial institution organizational behavior interviews. These interviews include specific questions regarding STEPP analysis and Porter's Five Forces.

In addition to addressing the interview questions, the review sought to provide an in-depth congruence model analysis for past and future change management programs that the Hawai'ian bank was sponsoring. STEPP and Porter's Five Forces analysis that was conducted was specific to the subject bank[115].

The questions and responses are included below. Relevant information gathered from the questions below have also been integrated and tied directly into the congruence model analysis.

---

[115]The bank's name is withheld due to a Non-Disclosure Agreement (NDA).

Analysis of Specific Interview Questions

Included below are the specific findings for the executive behavior interviews. Responses were combined by the interviewees or in some limited cases they declined to answer specific questions.

Human Resources

An interview by Senior Vice President of Human Resources provided the basis for the responses to the human resources (HR) questions. There are some noted instances where branch managers, the Senior VP's of other decisions provided answers that were specific to their areas of expertise.

Human Resources Questions and Answers

Human resources are at the center of an organization's operations. The headlines regarding discrimination against women and minorities in the workplace make this study even more imperative.

First set of questions

Included below is the complete interview, which providing valuable insights into the institution's corporate culture.

What is your company's employee composition?

How many employees?

There are approximately twenty-five hundred (2,100 +/) employees within the corporation's Hawai'i, Guam, American Samoa, Saipan/Tinian and Palau locations. Employees are situated primarily on O'ahu. This figure is down from 2008 when the number was approximately 2,500.

Per location?

The bank was not able to provide this figure at that time. They do not compile this information and it is an area that they could improve upon.

Diversity?
Gender (Male/Female)?

Age?

Ethnicity?

Years of service?

Diversity, gender, age, ethnicity and years of service are representative of national averages. The bank prepares an affirmative action report as a federal contractor to validate these results. Employees range in duration from one (1) day to fifty-two (52) years, with certain benefits being more important to specific age groups. The employee had been with the institution for fifty-two years (52) and was named Bert. According to HR this person was a pleasure to work with.

Organization chart?

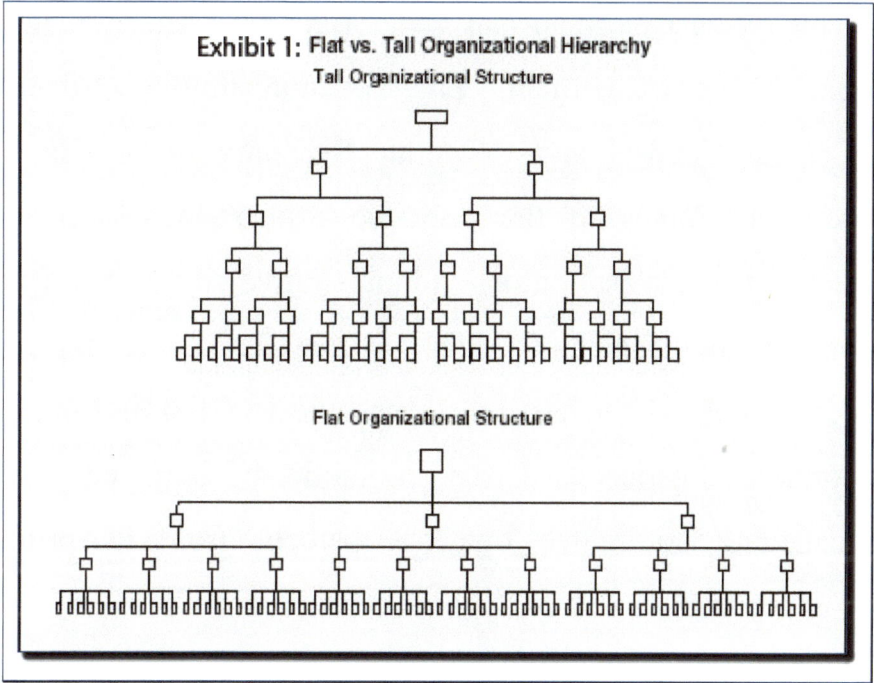

Exhibit 1: Flat vs. Tall Organizational Hierarchy

Tall Organizational Structure

Flat Organizational Structure

Bank provided org chart. Bank name withheld.

The organizational structure is fairly typical with a Chairman, Chief Executive Officer and President of the Bank, Vice Chairman and Chief Banking Officer of the Bank, Vice Chairman, Chief Administrative Officer, General Counsel and Corporate Secretary of the Bank being examples of the principle officers. One unique element to the institutions' administrative structure is the promotion of Senior VP's to the Bank's managing committee, which has direct exposure to CEO and other corporate officers. These promotions increased the managing committee by 25% percent from its prior composition.

General supervisory structure of other bank management includes Executive Vice Presidents, Senior Vice Presidents and Associate Vice Presidents, which are organized by corporate functionality (Retail Banking, Commercial Banking, Investment Services and Treasury – an aggregate of HR and other support functions).

Types of entrance testing? What are pros, cons?

Entrance testing is non-existent within the financial institution. This is primarily due to the inability to provide testing validation of entrance testing; or to benchmark it. For the Bank there are administration issues (logistical), legal issues, cost and relative questions associated with entrance testing that have led the institution to not pursue this avenue of pre-employment review.

Drug testing does exist at the Bank. It has the direct effect of further limiting the potential employee pool because the marketplace [banking and general employment] is so tight to begin with and unemployment even in the 2007-2008 recessionary conditions was so small. This is also the case in 2017-2018 with unemployment being at historic lows.

Do you hire from your competitor(s)?

The firm does hire from its competitors. This occurs because the marketplace and talent group for banking is so small. It also occurs because unemployment in Hawai'i is so low.

Discuss retail sales employee turnover concerns.

What does the business do to prevent turnover (if even necessary)?

The institution tries to hire, develop and reward talent in order to prevent employee turnover. HR doesn't view all employee turnovers as a negative factor. Some employees who resign or are terminated are not viewed as a good fit within the organization. In other cases they are not suited to the performance based sales position for which they are hired. In a second interview the Senior Vice President & Retail Sales Manager, Mortgage Banking Division at the Bank confirmed this.

Communication is utilized heavily to constantly encourage and steer employees in the right direction.

How are incentives and/or policies structured to address this challenge?

The Bank utilizes the Pathways to Leadership Excellence as the primary developmental program to build leadership talent for the future. It is also used to retain existing talent.

What is the average length of tenure in your front line retail sales positions?

Retail sales include mortgage and loan officers and depending on their skill sets they may be very successful or not at all. HR resources and the Mortgage banking group confirmed this information.

Tellers and other front line employees have the highest level of turnover. Talented tellers and other personnel are promoted to higher positions to retain them. They are also rewarded with incentives to stay in their existing positions.

Are there flex / part-time schedules available?

There are limited circumstances where flex or part-time schedules for employees are allowed. The majority of employees work a fixed schedule, nine (9) to five (5) or a similar combination of hours. This varies with the mortgage group whose employment is a lifestyle with meetings with realtors and loan originators during the day and paperwork to support these meetings in the evenings.

Do employees ever leave your bank to work at your competitor(s)?

This is a constant problem for the financial institutions because the marketplace is so tight. SVP participating in the interview indicated that in many cases the firm becomes a training ground for other competitors.

How is employee performance measured and compensated? How do you think this affects behavior in the organization?

360 evaluations[116] for what types of positions (if applicable to the corporation)?

---

[116]360-degree feedback is defined as follows: It is a tool that

The Bank doesn't conduct 360 evaluations.

Different structures for different positions?

There are definitely different pay structures for different positions. Compensation based programs utilize a base-pay rate (very low), plus sales quotas and goals.

Pay on quota?

---

provides employees with the opportunity to receive performance feedback from his or her supervisor or vice-versa. Each individual in a self-assessment also responds to many 360-degree feedback tools.

The VP interviewed indicated that his Mortgage Banking group stack ranks their compensation-based employees. The institution indicated that variable based compensation programs company-wide are incentive-based programs that are linked to the goals of a specific business (Mortgage or small business banking) and/or division. There are different incentive-based programs based on different divisions. Some of the programs are uncapped and are unlimited in their potential for incentive-based rewards.

Performance standards?

The business does have performance standard. Performance standards vary depending on the type of position, with support functions being evaluated differently that quota or compensation-based sales employees. Employees are generally scored on a 1 to 5 basis (Exceeds, Meets, Below, et. al.) for annual performance reviews.

Sales personnel mortgage and loan officers are also stack-ranked according to the VP interviewed. Those sales associates with the highest level of sales are ranked the highest; those with the lowest sales are ranked the lowest.

Annual review?

Annual performance reviews are conducted with a typical score range – Exceeds, Meets, Below, et. al. The scoring is based on a 1 to 5 scale ranking system.

How do you deal with healthcare issues? Do you have any special programs to promote employee health and wellness? If applicable, have you seen a return on investment?

The company does conduct wellness programs, which include the branded WeightWatchers®, brown bag lunches and health fairs. The Bank doesn't track the return on investment that any of these programs provide. This lack of qualitative or quantitative tracking of these programs is an area of improvement for HR and the institution's operations.

Does your corporation have a formal statement of core values? If not, what are the understood core values? What is the Mission and Vision statement? Do the core values, mission, and vision have an effect on behavior, and how so?

The corporation does have a formal statement of core values.[117] It is paraphrased as follows: Exceptional people building value for their clients, our communities, our shareholders, and the team.[118] The core values are tied into incentive based programs at the bank and have a direct impact on employee behavior as a consequence.

---

[117]Core values statement withheld for confidentiality reasons.
[118]Confidential Study. 10-K for 12/31/07. SEC Info. Filed 2/5/08.

What has been your most significant human resource challenges over the last few years and what is done to address them?

The HR department has had to deal with four (4) generations in the workplace. This is one of the greatest challenges for this group because they have had to design HR policies that meet a diverse group of people's needs. This has been very difficult for the bank because different types of benefits are more important to certain groups [generations] within the bank.

Retaining talented employees has been the other significant challenge due to strong competition and low unemployment.

In both instances HR has dealt with these issues by developing effective human resources policies and communicating as frequently as possible to ensure everything is functioning smoothly. Promoting and rewarding talented people with incentives are another mechanism that the institution utilizes to address generational gaps and employee turnover.

Corporate Culture

Discuss the Bank's corporate culture:

Degree of innovation; risk-taking

Attention to detail

Results focus

People focus

Team focus

Degree of competitiveness (among employee's)

Degree of stability, status quo

The degree of risk taking is higher for variable based or compensation based employees who are more likely to "ask for forgiveness", versus compliance or support positions who "ask for permission." All of the values stated above are critical to success at the organization.

Do you make hiring decisions based on organizational cultural fit?

The Bank makes hiring decisions based on organizational cultural fit. In general employees need to be results and team oriented, embrace change and communicate openly and honestly. The firm also hires for attitude and seeks people who are positive go-getters. HR hires based on specific skill sets, which is conducted on a case-by-case basis. Mortgage and small business banking hire more dynamic, assertive and proactive individuals. They also need to like people and have an innate ability to earn relationships and trust. For the variable pay plan originators people are hired on the understanding that this type of work is a lifestyle requiring meetings with realtors and branch managers during the day and paperwork in the evening. It also requires that these personnel be on call on iPhone's and other electronic devices to talk to their clients at all times of the day. It also requires that they travel to and from real estate locations.

Specific divisions or groups like mortgage banking and small business lending have more specific goals. The SVP noted that mortgage banking seeks employees who are fast (30 days to close when it is needed), flexible (power of the portfolio) and friendly (ninety-two 92% likely to recommend to friends and family).

Are there subcultures within the corporation? If so, what do you think drives the varying cultures? Do the groups get along with one another?

Subcultures are limited within the business because of the overarching theme of teamwork and incentive based programs that tie competition divisions back together.

How do your various programs affect the corporate culture? As applicable: Employee Stock Owner Plan (ESOP), bonuses and employee retention.

The company doesn't utilize an ESOP plan. It utilizes variable pay and other incentive programs to encourage and promote employee retention.

Who sets the tone for the predominant culture at the firm? What management behaviors, employee activities contribute to culture?

Management sets the predominant tone in the business, particularly the executive officers and managing committee. However, they are not ultimatum directives on the part of CEO and the other principle officers. Their "voices are sometimes louder than others" in terms of direction.

What aspect of your corporate culture needs to be changed (improved) and why? Are there any significant events or conditions that have negatively impacted culture over the last few years?

None of the personnel provided any specific examples of areas of where corporate culture can be improved. It may be inferred from the HR interviews and generalized statements that talent recruitment (i.e. the idea that the financial institution can never have enough talent) and employee retention are areas where the bank can improve its corporate culture to make the organization so attractive that recruitment efforts decline and retention levels increase. Corporate culture and incentive programs can constantly be improved to address this.

No specific examples of conditions or events were mentioned that have negatively impacted the firm's culture over the last few years. It is possible to conclude that the necessity of bringing a corporate turnaround specialist as CEO in the bank's past (November of 2000) divestiture of Asian and California branches, the removal of old employees and the recruitment of new executives to complete this work had a significant negative impact on corporate culture.

Small Business Banking Division

What have been your most significant Small Business lending challenges over the last few years and what is done to address them?

According to an interview with Vice President and Marketing Manager of the Bank, the most significant Small Business group's lending challenges include a slowdown of the U.S. and Hawai'ian economy and significant competition from other competitors.

How do you screen small business to prevent default on loans?

It is not appropriate to screen out potential clients. Instead the loan originators and underwriters through the application qualify or disqualify businesses. One disqualifying trigger occurs if a small business's debt-to-asset ratio is too high for real estate loans or other products and services.

Is your business division's organizational culture different in any way from the firm's culture at large?

The small business services group seeks employees who can be part of an integrated team. It is easy to train people to be bankers, but it is not as easy to learn to become a good team member.

Mortgage Banking Division

What have been your most significant Mortgage lending challenges over the last few years and what is done to address them?

This question wasn't specifically addressed. However, it was implied that the cooling of the real estate market has made it more difficult for mortgage and loan originators to complete real estate transactions.

How do you screen potential mortgage clients to prevent default on loans?

The corporation doesn't screen mortgage clients. This is not permitted. Instead the loan originators and underwriters may deny a mortgage application based on a lower FICA (Fair Isaacs Corporation) score, a debt to income ratio that is too high or not enough equity.

Is your business division's organizational culture different in any way from the institutional culture at large?

Overall culture is team based, but the mortgage-banking group tends to be more competitive. Employees are paid a commission and are variable-based salaries. If employees do well they make a very nice living. If employees don't do well they don't last very long. The culture of this division is very Darwinist in nature. However, it is brought back to the center and collective cooperation by the overarching incentive programs that reward employees for their performance.

Teller and Concierge Management

How has the management of tellers and other employees that deal with customers on a day to day basis changed over the last five (5) years? Ten (10) years?

The SVP being interviewed noted that tellers are trained to listen, recognize opportunities to promote financial products including mortgages, checking and savings accounts, retirement accounts and other investment services and products.

Many banking positions have become commission based. That represents a shift in the employment structure due to heavy competition.

What sorts of incentives and/or disincentives are in place to prevent tellers from making mistakes (withdrawals or deposits)?

Tellers are allowed a minimum number of mistakes.

How do you deal with difficult customers that enter the Bank on the ground floor in each of your branches?

An interview with the Branch Manager of the Kapahulu Branch on O'ahu indicated that he first tries to resolve any problems that a customer may be having. If the customer is mentally unstable or erratic he calls security.

What sorts of security measures are in place to deal with difficult customers?

Ample security personnel and cameras are available to survey the lobbies and other areas of each branch.

Is your business division's managerial culture different in any way from the culture at large?

The branch manager notes that his branch in Kapahulu is a closely-knit group that he can rely on to run the branch when he is absent (sick) or on vacation.

Questions for HR, Finance and Management

For the Porters Five (Market) Forces (an organizational/industry analysis tool) the bank was asked to share their insights on these topics.

Supplier Power: Relationships with your suppliers? Outsourcing (Customer Service)?

HR indicated that they have conducted limited outsourcing. The firm retains its call and support center in Hawai'i. Mortgage banking's SVP noted that his group has outsourced loan origination services through a Florida vendor's MortgageFlex Systems Inc.®

Buyer Power: Relationship with your customers: Retail, Wholesale (if applicable), Trust management, Commercial and Industrial?

No formal discussion was provided on this as it relates to the Porter's Five Forces principles. However, indirectly the Bank is heavily focused on the customer and their satisfaction and works to create a uniform brand with a suite of products to create one-stop-shop for customers. The SVP interviewed notes that the customer is critical to the bank's success. It may be inferred that on an individual level the customer has limited power, but that because the firm treats the customer as an aggregate group they have more power when defined this way.

Barriers to Entry: What strategies do you employ to keep barriers as high as possible? What assets does the firm possess that differentiates it from potential competitors

The Bank doesn't focus on barriers to entry per se. Instead the corporation focuses on its own brand and customers to ensure that the products and services stand alone; and that this acts as a "natural barrier" that differentiates the institution from the competition.

Threat of Substitutes: How do you address threat of substitutes? Product differentiation?

The organization does track what the competitor is doing. However, the firm is again more focused on its own brand and customers to ensure that the products and services stand-alone. This works to prevent customers from leaving the Bank for the competition.

Degree of Rivalry: How do you address industry concentration, difficulty of exiting the market, switching costs (customer moving from one bank to another)?

This is a natural process that evolves and there is a certain degree of churn that occurs in the industry. The key is the firms' focus on providing the highest quality brand and service to their customers to address the high concentration of banks/financial providers in Hawai'i.

For the STEPP market forces: social; technological; economic; political and/or legal; and physical (location). Please share your insights on how any dynamic within one or more of these forces drive your business strategy. What changes have been made, or will be made, in response to any developments or concerns within any one of these forces?

The firm's human resources were the only division that emphasized that technology has changed the manner in which payroll and other benefits are accessed through on-line open enrolment. The business utilizes BlueFlash® for its open enrolment benefits for health care on at least an annual basis. Employees can view their pay stubs and other HR materials on-line in addition to the open enrolment program. Technology was the only STEPP force that any of the divisions specifically mentioned as having an impact of the organizational behavior and business strategy of the Bank.

Have you ever considered outsourcing customer service or other divisions (if you do not already do so) outside of Hawai'i? Outside of the U.S.? What are the pros and cons?

The company has kept its call center and customer support functions out in its Kapolei facility on the Leeward side of O'ahu. The corporation believes that strong customer service is critical to maintaining the strength of its brand and overall efficiency. Its core customer base is located in the Hawai'ian Islands and customer service representatives that are hired from this marketplace create the strongest buy in to the institution's brand.

What functions do you outsource now? Marketing? Logistics? Data processing?

HR indicated that they have conducted limited outsourcing. The Bank retains its call and support center in Hawai'i. The SVP interviewed noted that his group (Mortgage banking) has outsourced loan origination services through a Florida vendor's MortgageFlex Systems Inc.®

How do Hawai'i, national (Mainland) or international practices and procedures compare with the firm?

The general banking practices and procedures of the firm are very similar to national or international ones. HR noted that recruiting, advertising and marketing is tailored specifically to the cultures in Guam, American Samoa, Palau and CNMI/Saipan and it differs from the same recruiting, advertising and marketing that is conducted in Hawai'i.

Why An Investigative Review of Banking Culture Is Important

After the interviewing process is completed the information can then be used to begin to map the organizational culture within the Bank, determine who the influencers are and the roadblocks. The end result of this process is an improvement in revenue and bottom line profitability for the institution.

It can also be used to remediate toxic culture and improve diversity initiatives within the organization. The results of this process lead to a healthier bank and better employees. It requires more than webinar training on diversity and regulatory requirements. It takes real work and involvement.

The results of the process can also be used to determine areas in which the Bank is not taking enough calculated risks and areas where this process can be improved, particularly in gradually branching out into private equity and venture capital deals, for both their wealth management and institutional clients.

Venture Capital Investments

The rise in private client, venture capital and other institutional investing can be a calculated risk that institutions like the subject Bank take to grow their businesses. It can also involve creating a separate LLC, LLP or corporate structure under the parent bank. Deals can be small and aligned with the Bank's mission initially, including fintech and blockchain; and then gradually take on deal flow that is outside of the institution's core mission to include: biotech, genomics, SAAS, and PAAS.

Organizational Culture and VC Funding

Organizational behavior analysis is equally as important to venture capital funding of accelerators and incubators. Typically many incubators and accelerators in the start-up ecosystem take a percentage of the founders' equity (at 10 to 15%) for $20,000 Convertible Note[119] at the low end of the threshold and up to $120,000 at the higher end of the spectrum. Most of these same VCs, accelerators and incubators (three month programs) offer paid classroom learning experiences where you can receive training on legal incorporation (typically California, Nevada and/or Delaware), accounting best practices and other strategies.

---

[119]The value of a Convertible Note is not set initially. After a company secures later investment the debt converts to equity based on a Seed or A-Round valuation.

There are ranges of startup conferences that allow companies to apply for a booth and/or the opportunity to present in front a panel of judges. What is not disclosed at the beginning of this process is that these companies typically require a fee to participate after you are accepted.[120] However, they only disclose the fees after the screening process is completed.

Other companies in California and New York charge for pitching practice and video production service. There are even some companies, particularly in California that charge people to present to investors. These pay to pitch companies disclose the fee structure after the first presentation is made by the presenting startups. Those that want to continue pay a fee of $2,500 to $5,000 to present to the next round of investors. At the end of this process these same investors take an equity stake in the start-up. In essence they charge to present and then own a part of your company. This double-dipping process is not typical in the VC community and is frowned upon in California and New York by investors.

---

[120]Confidential Case Study. Corporate Names Withheld.

Problems Early Financing Creates for Startups Culture Later

The pay to play nature of early start-up culture creates problems for an enterprise's corporate culture later. For a founding team to participate they must be wealthy to begin with. Low- and moderate-income founding teams wouldn't be able to pay for rent in San Francisco with a $20,000 convertible note.

Why is this the case?

The current economic reality of this situation is as follows: Rents are $4,000 to $5,000 a month in San Francisco and New York. This means that a non-wealthy founder would be able to purchase three months of rent (at $5,000 a month, plus security deposit) in what is typically a three-month program. Now these founders are indebted to investors and they have a shortfall in housing rent having spent nothing on technology development or marketing. This requires them to immediately seek funding again to cover their expenses.

For a founder in a accelerator that offers $120,000 in equity funding this pays for a single engineer, leaving no money for housing, legal fees and/or other start-up issues. In this situation the founding team has to be raising funds instead of focusing on product development.

What does this result in?

Independently wealthy individuals that can afford to weather any short-term economic downturn found most companies. This creates a huge organizational culture issue from the onset. Many low-income or minority founders are eliminated from the process before they can even start. This creates significant issues with diversity and women founders that are baked into the process from the onset of a corporation's founding. It is one of the primary reasons that startup culture has been highly criticized for a lack of inclusion and problematic issues with harassment and discrimination.

Improving Diversity and Inclusion in Startup Culture

Investors want to receive a return for their investment. There is nothing wrong with this. However, in order to create a more inclusive startup community, investors must back women and minorities and take on slightly more risky investments. By doing this they can build diversity into startup culture from the beginning.

Creating an Environment Where Fear, Uncertainty and Mistakes Are Tools

It is critical that start-ups create a culture where fear, uncertainty and mistakes are not punished. They are learning tools providing an opportunity for improving processes. They should not be weapons that are used to punish and ridicule employees. The news headlines in 2018 reflect the actual reality: bullying, harassment and the worst schoolyard behavior of many managers. Many of these cultures were so extreme that managers were engaging in criminal conduct, far worse than any college fraternity. The targets of this conduct: women and minorities (or both).

The juxtaposition of federal labor law and the advertised PR campaigns of these companies could not be more different from the actual reality of these workplaces. The convert uncertainty and fear into actual weaponized strategies to harm and demoralize employees.

Fear and uncertainty should be positive agents for change. Not tools of manipulation and power designed to undermine and weaken an employment environment.

Lessons Learned From A Dysfunctional Banking Culture

At a U.S. bank based in the Western United States, examining a change management process that resulted from a dysfunctional culture is critical.[121] The lessons learned from an improperly functioning banking culture are just as important as those found in a highly efficient operation.

IT Dysfunction and Process Improvements

The CEO of a major bank called the IT help department and identified himself by name. The IT Department was based in a large-corporate office building, in a separate physical location from the headquarters. The IT staff transferred the CEO's call and after several forwarded calls he was asked: "Are you willing to be placed on hold?" He responded, "Yes." He then waited patiently for 45 minutes and never said a word.

---

[121]Confidential Case Study. Corporate Names Withheld.

The IT Department made their own CEO, an individual in charge of a corporation with thousands of employees, wait 45 minutes to resolve problems with his computer. After the 45 minutes of time was up this CEO thanked the IT Department for providing him with assistance for his problem. He then hung up the phone. He had waited patiently all this time and had cancelled other meetings to take care of a problem with his computer's operation before taking a flight to New York City.

The IT Department was so focused on their policies & procedures and infighting (transferring the call from person to person) that they lost sight of their role in the organization: to serve their customers and clients. In this case they allowed their narrow scope of vision to cloud their work: it was their direct responsibility to serve their clients, in this case their fellow employees, regardless of their personal feelings to one another and/or towards the CEO.

Unbeknownst to these IT department employees' changes were about to me made. The CEO took his NYC itinerary from his team and was driven to the San Francisco airport.

3 months later, the IT Department was shocked to learn that most of their department was being transferred to a new operation in the mid-West in a brand new facility. Their work was going to be completed by an outsourced 3rd party. Employees would be given the option to relocate from California to this new location over a 6-month period. If they didn't elect to move they would receive a severance package from the bank based on years of experience. Most employees (due to their families being based in California) elected to receive the severance package.

The most striking aspect of this process was the conversations that I overheard and/or was directly a part of during this 6-month timeframe. The impacted employees couldn't believe that they this was happening to them.

My wait in the same IT Department was as follows: 1-month and twelve days to receive a laptop for work. The delay was comical; the IT department was unable to set-up a simple laptop once it arrived in California after a month delay for another two-weeks. I took notes on a notepad for this entire period while my machine was configured. The emails as a part of a new team had piled up for this lengthy duration.

When the new outsourcing group took over the problems where addressed immediately over the phone; or in exceptional cases – computers where shipped via FedEx back to the mid-West location. This newly implemented procedure was highly efficient.

From an organizational behavior analysis perspective this change was absolutely critical. The roadblocks and infighting between employees had become insurmountable. The only remedy in this case was to remove this portion of the IT Department. It had become easier to go to Best Buy® or Target® and buy a laptop directly. Configuring the security protocols, email and remote firewall access was only part of this process that couldn't be completed directly due to administrative login protocols. The backlog of IT Department tickets immediately was immediately cleared away and productivity was vastly improved.

Conclusions

The lesson in this process is simple: Never let yourself or your employees become so narrowly focused that they lose site of the long-term strategic goals of customer service. The consequences as outlined in this case are typical: blinders and auditory blocks and other sensory perceptivity that fail to provide periphery insight into the risks surfacing outside the range of vision, hearing and tactical range.

Toxic Behavior and Corporate Cultural Behavior

However, the institution still contained many hostile and retaliatory aspects within its organizational culture despite the procedural IT changes.

After the IT modifications were made it was not uncommon to still hear slurs against other employees in the halls or offices. In several instances I watched supervisors extend obscene gestures to each other.

What are some of the lessons learned from this type of caustic employee behavior?

Never allow this philosophy to be replicated in any culture that you start, are hired to run and/or advise.

Conclusions

In situations that are unreasonably targeting protected employment classes and organizationally dysfunctional the best thing an employee can do is *walk away*. Unfortunately, for many negatively impacted employees their departure pays for rest of their fellow 401(k) team members plan fees. In essence they are paying twice for the contaminated culture.

The problems in these types of companies are so deep seeded and systemic that it is impossible to cut the rotting and fermenting aspects of this culture out without killing the patient at large (i.e. the firm itself).

However, founders and managers should not allow this to deter their enthusiasm for starting, building or being hired to turn around a struggling company. It starts with the individual. Focus on implementing the following:

1. Encouraging employees to explore their personal and professional development;
2. Inspire physical fitness during and outside of work (whether you play golf, walk, run, cycle, swim, ski, snowboard, surf, et. al.);

3. Promote and reward experience from the law enforcement, intelligence, military and veteran communities (it adds a valuable alternative perspective to the workplace);

4. From an investor perspective it requires thorough due diligence and targeting companies that are aligned with your personal philosophy (online merchants who treat their warehouse employees like a critical part of their corporate family and team not expendable and disposable assets to replaced on a seconds notice and whim);

5. It requires identifying and correcting toxic institutional culture before it becomes systemic and untreatable; and

6. Mistakes should be allowed (they lead to new technology, chemistry, nanotechnology and processes).

The future starts now.

It starts with making individual changes and leading by example. It begins by conducting site visits of regional offices, the IT department, the factory floor, and the outsourced call-centers.

It is initiated by meeting with employees at every level. It even includes joining fellow employees on the factory floor to learn about how they work and take feedback on things that aren't working for them.

Many hedge fund managers that I have spoken with that have real estate (as an example) in their portfolio have limited knowledge of the following:

1. What their apartment complex is like;
2. How it is constructed;
3. The types of materials that it is built with; and
4. What their portfolio's strengthens and weaknesses are.

They have never visited the investments that their portfolio is representing. If a housing discrimination, ADA compliance an/or hazardous materials issue emerges with one of their properties they are completely in the dark. This is a very naïve and foolish management style.

The same types of issues are present with other asset classes. It is imperative to know details about what you are investing in, not just financial numbers and investor prospective documentation.

Physical Audits

ExxonMobile's® *1. Safety and Operations Integrity* summary highlights the constant and imperative to dig deeper and examine at granular level when an operational safety issue spike occurs.

In 2011, pursuant to Exxon Mobile's® notice of 2016 Annual Meeting and Proxy Statement[122], the company provides a visualization that graphically displays that although ExxonMobile's® safety record falls below the industry benchmark it experienced a spike in safety incidents in that year.

---

[122]Annual Meeting & Proxy Statement. ExxonMobile®. 4/13/16.

From an investor's perspective it is important to know & ask why? Did a major or explosion or accident occur at their operations and why? The spike in data, however small, requires further due diligence. Why is this? Safety incidents result in regulatory scrutiny from OSHA and other international safety regulators.

In the emerging start-up world and asset management industry it is particularly important to address proactively before they become an issue that attracts media and/or investor attention.

Knowing your assets inside and outside can prevent a manager from being caught off guard when a problem emerges, whether it has do with the structural integrity of housing, renewables infrastructure (wind and solar farms), real estate; a medical or wearable safety issue; a data privacy or cybersecurity issue; or a human resources and/or organizational culture problem (harassment against female, minority or other protected classes).

Conclusions

Don't be caught off guard as a manager. Know your investments at a detailed and macro level. If your company has hidden issues that a physical audit reveals, take proactive steps to remediate directly. If your portfolio is liquid (not illiquid via a hedge fund), remove these risky assets as soon as possible. Systemic issues in a start-up or corporation eventually surface and cause major structural issues, like dry-rot in the windows, roof lines and doors of a house, leading to electrical fires and/or major repairs. Inspect your assets frequently. If you can't travel to examine them, hire an auditor and/or consultants to generate reporting for you. These types of audits, in addition to the financial and legal due diligence will protect you in the long-term, resulting in a strong portfolio, a solid reputation and strong returns for your shareholders.

Overarching Themes

It is critical to constantly evaluate and re-assess mission critical organizational behavior issues before they become a newsworthy headline (lack of diversity, toxic culture, et. al.) Technology is secondary in many respects to addressing these issues. Programming is very procedural: coders regardless of their race or gender can achieve similar outcomes.

From a management perspective it is imperative that you know the inner workings of your investment portfolio: whether it is in the apparel, software, restaurant or housing business. Take the time: Travel to the factories, tour the properties and test the clothing.

Larger societal issues are a huge market opportunity. Innovations in energy, transportation and housing are desperately needed, particularly in material invention. In order for science and technology to be truly groundbreaking it must begin to tackle the really difficult issues.

In the medical arena engineers can and should address larger diseases and ailments: HIV, MS, Parkinson's disease. For technology to be truly disruptive this has to happen. It is not.

www.ingramcontent.com/pod-product-compliance
Lightning Source LLC
Chambersburg PA
CBHW041305210326
41598CB00011B/851